A JOURNAL-TABLE of our *Voyage*, in the Ship *Marquis*, the Latitude of 23 Deg. 10 Min. *North*; to the Island *Guam*, in the Latitude of 13 Deg Year 1709.

Months and Days.	Course corrected.	Diste. sail'd.	Northings in Miles and Tenths.		Southings in Miles and Tenths.		Eastings in Miles and Tenths.		Westings in Miles and Tenths.		Latitude per Observation.		Latitude per Estimation.		Longitude.	
		Miles	Miles.	Tenths	Miles.	Tenths.	Miles.	Tenths.	Miles.	Tenths.	Deg.	Min	Deg.	Min	D.	M
Jan. 11,12,13	S.24 Deg. W.	106	0	0	97	0	0	0	42	5	21	35	21	33	0	4
14,15	S. S. W.	115	0	0	106	3	0	0	44	0	19	47	19	47	1	2
16,17	S.S.W. ¼ W.	116	0	0	102	0	0	0	55	0	18	5	18	5	2	2
18,19	S.W. ½ S.	120	0	0	93	0	0	0	76	0	16	32	16	32	3	4
20,21	S.W. ¼ S.	146	0	0	92	0	0	0	113	0	15	0	15	0	5	4
22,23	W.6 Deg.S.	236	0	0	28	2	0	0	234	2	14	30	14	30	9	4
24,25	W. ½ S.	250	0	0	24	5	0	0	248	8	0	0	14	6	14	
26,27	W.	206	0	0	0	0	0	0	206	0	13	36	13	36	17	2
28,29	W.4 Deg. S.	280	0	0	12	0	0	0	279	0	13	25	13	24	20	2
30,31	W.	260	0	0	0	0	0	0	260	0	13	25	13	25	25	
Febru. 1, 2	W. ½ N.	280	14	0	0	0	0	0	279	0	13	36	13	39	29	4
3, 4	W.2 Deg.S.	290	0	0	10	0	0	0	275	0	13	29	13	29	34	2
5, 6	W.	240	0	0	0	0	0	0	240	0	13	29	13	29	38	3
7, 8	W. 2 Deg S.	281	0	0	9	0	0	0	280	0	13	20	13	20	43	2
9,10	W. ½ N.	251	25	0	0	0	0	0	249	0	13	45	13	45	47	4
11,12	W.2 Deg.S.	270	0	0	9	4	0	0	269	7	13	37	13	37	52	1
13,14	W.2 Deg.N.	270	9	7	0	0	0	0	269	7	13	47	13	47	56	3
15,16	W.	230	0	0	0	0	0	0	230	0	13	47	13	47	60	3
17,18	W.	230	0	0	0	0	0	0	230	0	13	47	13	47	64	2
19,20	W.4 Deg. S.	271	0	0	19	0	0	0	270	0	13	28	13	28	69	
21,22	W. ½ S.	212	0	0	10	0	0	0	210	0	0	0	13	8	72	3
23,24	W.	180	0	0	0	0	0	0	160	0	13	8	13	8	75	2
25,26	W.	195	0	0	0	0	0	0	195	0	13	8	13	8	78	4
27,28	W. ¼ S.	200	0	0	20	0	0	0	199	0	12	48	12	48	83	
March 1, 2	W. by N.	240	33	0	0	0	0	0	237	4	13	23	13	23	87	4
3, 4	W.	290	0	0	0	0	0	0	290	0	13	23	13	23	90	5
5, 6	W.	115	0	0	0	0	0	0	115	0	13	23	13	23	92	5
7, 8	W.	176	0	0	0	0	0	0	175	0	13	23	13	23	95	4
9,10	W. 2 Deg.N.	210	7	4	0	0	0	0	209	0	13	30	13	30	99	3
11	W.	47	0	0	0	0	0	0	47	0	13	30	13	30	100	2

This Day at Noon, the *South* Part of the Island *Guam*, bore *West* and by *South*, Distance 8 Leagues; and the *Northermost* Part *North East* an

A VOYAGE TO THE South Sea, AND Round the World.

Perform'd in the
YEARS 1708, 1709, 1710, and 1711.
BY THE
Ships *Duke* and *Dutchess* of *Bristol.*
BEING
A CONTINUATION of the Voyage from *California*, through *India*, and *North* about into *England*.

The DESCRIPTION of all the *American* Coasts along the *South* Sea, with above 300 Bearings of the Land, the principal Harbours, and three large Charts, all taken from the *Spanish* original Draughts, never before printed.

With a TABLE of the Latitudes and Longitudes of all Places, from *California*, to the Streights of *Magellan*.

To which is prefix'd,
An INTRODUCTION, wherein, besides other material Particulars, is an Account of the Cargo of the *Acapulco* Prize, of the Commodities the *West Indies* are furnish'd with, by way of Trade, from the several Parts of *Europe*, and what Returns come from thence.

Vol. II. and last.

By Capt. EDWARD COOKE.

LONDON, Printed by *H. M.* for B. LINTOT and R. GOSLING in *Fleet-Street*, A. BETTESWORTH on *London-Bridge*, and W. INNYS in *St. Paul's Church-Yard*. MDCCXII.

X Defoe. 711. C77V v.2

TO THE

Right Honourable

ROBERT,

EARL OF

Oxford and *Mortimer*, Lord High Treasurer of *Great Britain*, &c.

This Second Part of the Voyage to the *South Sea*, and round the World, is most humbly dedicated, by

EDWARD COOKE.

THE CONTENTS OF THE VOYAGE.

CHAP. I. *Departure from California; the long run across the South Sea, to the Islands* Ladrones; *Arrival at* Guam, *one of that Number; courteous Entertainment there by the* Spaniards; *Letters and Certificates on both Sides; Variation in those Parts,* &c. p. 1.

Chap. II. *A brief Account of the* Marian *Islands, commonly call'd* Ladrones; *Description of the Island* Guam, *or* Iguana; *the Islands of* Solomon; Paraos, *a Sort of Boats; the* Rima, Ducdu, Areca, *and Pine-Appple Fruits,* &c. 13.

Chap. III. *Departure from the Island* Guam, *or* Iguana; *see some small Islands; an Account of Spouts;* Moratay *and* Gilolo *Islands; of the Monsons, Signals, dread-*

A 3 *ful*

The CONTENTS.

ful Weather; Mindanao, *and the* Philippine *Islands*; *their Trade.* 23.

Chap. IV. *The Voyage continu'd among the Islands of* India, *and through several Streights, to* Batavia, *the Capital of the* Dutch *Dominions in those Parts*; *some Particulars of the Islands* Bouro, Cambava, Wanthut, Buton, Solayo, Madure, Carimon Java, *and the General's Island.* 33.

Chap. V. *How a Day is gain'd or lost, in sailing round the Globe*; *and short Account of the Road and City of* Batavia; *victualling and refitting there*; *Distribution of Plunder*; *Money advanc'd to Officers*; *the Ship* Marquis *sold*; *Orders and Resolutions of the Committee.* 52.

Chap. VI. *The Passage from* Batavia, *to the Cape of* Good Hope; *the Journal Table*; *Letter from thence to the Owners*; *Description of the Cape, the Town, and Natives of the Country about it*; *Preparations to return into* Europe. 62.

Chap. VII. *The* Dutch *Admiral's Sailing Orders.* 75.

Chap. VIII. *Departure from the Cape of* Good Hope; *Islands of St.* Helena, *and the* Ascension *in the* South; Bora *and* Shetland *Islands, in the* Northern *Sea*; *Arrival in* Holland, *and what happen'd there*; *Departure thence, and Arrival,*

first

The CONTENTS.

first at the Downs, *and then in the* Thames. 91.

Contents of the Description.

CHAP. I. *The Sea-Coasts,* &c. *from the City of* Panama, *on the* Isthmus *of* America, *to* Callao, *which is the Port to the City of* Lima, *Capital of* Peru.
110.

Chap. II. *The Sea-Coasts,* &c. *from the Port of* Callao, *in the Kingdom of* Peru, *to those of* Caralmapo *and* Chiloe, *the most Southern in* Chile. 199.

Chap. III. *The Sea-Coasts,* &c. *from the Port of* Panama, *on the Isthmus of* America, *to that of* Acapulco, *in the Kingdom of* New Spain, *and thence to* California. 257.

Chap. IV. *Of the Winds and Currents in the South Sea; as also a large Table of the Latitudes and Longitudes of all remarkable Places along that Coast.* 313.

Directions to the Bookbinder.

Place the first Journal-Table at Page 3.

ERRATA.

PAGE 4. Line 21. for *Stern*, read *Stem*.
Pag. 12. l. 6. f. *Prettana*, r. *Rattena*.
Pag. 18. l. 1. dele *other*.
Pag. 43. l. 33. f. *Pury*, r. *Jury*.

THE
Introduction.

AN Introduction to this second Volume, as well as to the first, may perhaps be carp'd at by those who read rather to find Faults, than for Information or Entertainment. The Omission of any Thing material, would be a more reasonable Objection, than this Method of inserting some Particulars, which could not so properly be brought into the Course of the Journal. A few Words at the Conclusion of the Introduction to the first Volume misunderstood, gave some Persons a sufficient Occasion, as they thought, to censure the Performance. It is there said, that a *continu'd Account of Winds, Latitudes, Longitudes, and such other Maritime Particulars, would be of little Use, and might prove heavy and tiresome*; whence, without any other sufficient

Ground

The Introduction.

Ground, the said Persons must conclude, that there had been an Omission of observing the necessary Latitudes and Longitudes. But the true Meaning of those Words, is no other, than that it was thought needless to tire the Reader with an exact Diary of Wind and Weather for three Years together, and the Latitudes and Longitudes during a long Run across the Ocean, where there was not any Land, nor so much as a Current to be known by those Observations; for wheresoever any Thing occurr'd, which might be ever observ'd by the Position, it is carefully set down; and to swell a Volume with what could neither be of Use, nor afford Entertainment, would have been altogether superfluous. This may suffice to satisfy the World, that nothing has been omitted, wherein the Reader might find his Profit or Satisfaction.

Others, who have had the Leisure and good Fortune to read some Translations of *Spanish* Histories relating to, or Travels into the *West Indies*, have complained, that the said first Volume is fill'd up with Collections, and some of them the same Things they have read before. Those Gentlemen will do well to call to Mind, that the most judicious Travellers have been applauded for giving us short Abridgments of the History of those

Coun-

Countries where they have travell'd; as to inftance but in one, though many others might be nam'd, has been done by the generally approv'd *Gemelli*, who has never been condemn'd for having told us many Things of *China*, *New Spain*, *Perfia*, and other Parts, which were before fufficiently known to all thofe who are converfant with Books of Travels, and Defcriptions of Countries. There was no new Difcovery made or intended in this Voyage to the *South Sea*, and confequently no Matters altogether ftrange and unheard of, could reafonably be expected. All *America* has before been treated of in every Language; and yet, without Vanity or Prefumption, it may be faid, there is fcarce fo much to be found elfewhere, in fo fmall a Compafs, as has been deliver'd in that Volume. All Perfons have not numerous Collections of Books, nor Time to read them; and it would have been extreamly difagreeable to a much greater Number, who being utterly unacquainted with the Hiftory and Defcription of *America*, yet may be induc d to read this new Voyage, fhould they find nothing in it, but tedious Runs at Sea, with only an Account of the Town of *Guayaquil*, and the taking of fome Prizes, and be left entirely in the Dark, as to all thofe wealthy

Coun-

Countries pafs'd by, and fcarce touch'd upon.

Thofe few who have thought fit to raife thefe, or any other as immaterial Objections, may perhaps, upon ferious Recollection, have been convinc'd of their Miftake, whilft the more confiderable Part of Mankind has afforded that firft Volume fuch a Reception, as leaves no Room to doubt of their Approbation. However, that Work did not appear Abroad as compleat; nothing in this World can pretend to Perfection; and the Promife then made of a fecond Volume, was Inducement enough to believe, that whatever might be wanting there, would be fupply'd in the next. This, there is Ground to hope, has been now perform'd.

The reft of the Voyage Home from *California*, takes up the firft Part, with exact Tables of the feveral Runs, from *America* a-crofs the Pacifick Ocean, to the Iflands *Ladrones*; from *Batavia* thro' the Sea of *India*, to the Cape of *Good Hope*; and from thence *North* about into *England*; and fatisfactory Accounts of all Places feen or touch'd at, or of any remarkable Accidents or Obfervations.

The fecond Part contains the Defcription and Bearings of all the Coafts from
Cali-

The Introduction.

California, to the Streights of *Magellan*, three large Cuts of all those Coasts, curiously drawn according to Art, and answering to the Distances in the Description, which many of the *Spanish* Draughts do not, being taken for the most Part only by the Eye, and consequently only serving to give a very imperfect Idea of what they represented. Here are also all the Harbours of any Note, all of them taken from the *Spanish* Draughts brought over this same Voyage. There is added an Account of the Winds and Currents in the *South Sea*, and a large Table of the Latitudes and Longitudes of all Places any Way remarkable along that whole Coast of *America*.

Some other Particulars, very necessary to be added to this Work, could not so properly be intermix'd with either of the Parts above-mention'd, and have been therefore reserv'd for this Introduction. These are, the Cargo of the *Acapulco* or *Manila* Ship, taken in the *South Sea*; a Particular of all the Commodities transported from every Part of *Europe*, to the *West Indies*, and of the Returns from thence, which at once will satisfy the Curious what is proper for the *South Sea* Trade, and what we may expect from thence; and lastly, a fuller Account of the Man found on the Island

John

vi *The Introduction.*

John Fernandes, in the *South Sea*, than we were able to give in the firſt Volume, being then preſs'd to publiſh it with all poſſible Speed, and having ſince receiv'd as much Information as is requiſite in an Affair of that Nature.

In the firſt Volume we gave the Particulars of the Cargo's of the *Duke*, *Dutcheſs*, and *Marquis*, being all that was found valuable, and worth preſerving, Aboard the ſeveral Prizes taken in the *South Sea*, and at the Town of *Guayaquil*. When the *Marquis* was ſold at *Batavia*, her Lading was there divided among the three remaining Ships, *Duke*, *Dutcheſs*, and *Batchelor*, as they then call'd the *Acapulco* Prize, of whoſe Lading we are now to ſpeak.

The

The Introduction.

The Cargo of the Acapulco *or* Manila *Ship, taken in the* South *Sea by the* Duke *and* Dutchess, *private Ships of War belonging to* Bristol, *and call'd the* Batchelor *Prize.*

	Pieces
Allejars	82
Atlasses	52
Bafts	188
Cottoneas	291
Calicoes colour'd	6603
Ditto white	4372
Counterpoints, divers sorts	206
Cossaes	270
Chints, divers sorts	24289
Chint Sashes	24
Chelloes	362
Charradorees	18
China flower'd Silks	5
Damasks	120
Dimities	460
Diapers	77
Elatches	3106
Fans	5806
Gurrahs	1180
Ginghams	263
Guinea Stuffs	235
Humhums	105
Handkerchiefs	

	Pieces.
Handkerchiefs Pieces	38
Ditto single	157
Long Cloth	2577
Mulmuls	55
Neck-cloths	123
Nillaes	580
Niccaneas	8020
Photees	152
Pelongs	1236
Paunches	16561
Palampores	4053
Petticoats	265
Quilts	14
Romols	548
Ribbons, divers sorts	6834
Ditto flower'd with Gold and Silver	481
Silk Stockings	4310 Pair.
Silk raw of *China*	28502 Pounds.
Ditto thrown	11990 Pounds.
Ditto sewing	1370 Pounds.
Ditto Bengal	61 Pounds.
Ditto sleve	6581 Pounds.
Ditto Fringes	194
Sooseys	115 Pieces.
Stockings Cotton	1084 Pair.
Sannoes	425 Pieces.
Sattins and Taffaties, divers sorts	7008 Pieces.
Ditto flower'd with Gold and Silver	192 Pieces.
Silks divers sorts	511 Pieces.

ilk

Silk Sashes	341
Ditto of Calico	544
Silk Gowns	37
Tanbes	454
Musk	5997 Ounc.
Cinnamon	9719 Pounds.
Cloves	1182 Pounds.
Benjamen	3300 Hund.
Bees Wax	152 Pounds.
Gum Elemia	120 Pounds.

China Ware, several Chests and Jars.
Several Parcels of odd Things.

At this Time, when a Trade to the *South Sea* is so much talk'd of, all Persons are desirous to know wherein that Trade will consist, and it may be expected something should be here said of it. The one Branch of it, which is the Commodities to be had from thence, seems to have been sufficiently answer'd in the first Volume, where, having treated of the Product of every Part, it naturally follows, that every Man, who knows what those Countries afford, may judge what is fit for our Use, and worth the bringing. However, to save some the Trouble of so much Reflection, and satisfy all Inclinations, all the Goods generally brought from the *West-Indies*, shall be here laid down together to be seen at one View, and with them the several Sorts every Country in *Europe* furnishes *America* with.

A compleat List of all Commodities transported from any Parts of Europe, *to the* Spanish West-Indies.

From FLANDERS.

Picotes, a Sort of Woollen Stuff.
Ditto half Silk.
Palometas, half Worsted.
Ditto half Thread, half Worsted.
Damasks all Worsted.
The same half Thread.
Lanillas white.
Ditto black.
Mix'd *Quinietas.*
Hollands,
Baracanes.
Womens woollen Hose of *Tournay.*
The same for Boys and Children.
Hair Chamlots of *Brussels.*
Lamparillas half Silk, half Worsted.
White Thread Lace.
Black Silk Lace.
Precillas, brown course Linnen.
Ditto white.
Bramantos brown.
Ditto white.
Ditto fine.
Hounscots of three, four, and five Seals.

Strip'd

Strip'd Linnen of *Gant*.
Gant Linnen fine.
The same of *Courtray*.
Damasks of Silk and Thread.
White Thread ordinary.
Ditto fine of few Numbers.
Ditto courser of many Numbers.
Thread of all Colours.
Thread Laces or Twists.
Cotton Ribbon.
White Filliting.
Red Tape.
Whip-cord large.
Ditto small.
Hair-buttons,
And several other sorts of Haberdashery.

From HOLLAND.

PEpper.
Cloves.
Cinnamon.
Nutmegs.
Serges in Grain.
Black *Leyden* Says.
Ditto of *Delfe*.
Fustians.
Broad *Hollands*.
Ditto narrow.
Strip'd Linnen.
Thread of all sorts.
Sail Cloth.
Cables and Rigging.

Ropes and Pack-thread.
Pitch and Tar.
Benjamen.
Motillas of Silk.
Ditto Wooll.
Borlones.
Ditto branch'd or flower'd for Quilts.
Velvets, and
Plushes.

From ENGLAND.

MIX'd Serges.
Long Ell-broad *Perpetuanas*.
Long Yards *ditto*.
Cheneys printed and water'd.
Silk Hose.
Colchester Bays, dy'd and white.
Worsted Hose fine.
Ditto second sort.
Woollen Hose for Men, Women, and Children.
Wrought Pewter.
Tin in Blocks.
Black Hounscot Says.
Ditto white.
Fustians.
Scotch Linnen.
Benjamen the second sort.
Lead.
Cloths broad.
Ditto narrow.
Scarlet Serges.

Calicoes

Calicoes dy'd.
Pepper.
Musk
Amber, and
Civet.

From FRANCE.

Velvets.
Brocades.
Sattins.
Roan Linnen.
Ditto Blancartes.
Ditto Florettes.
Ditto Cofres.
Cambricks.
Kenting broad.
Ditto narrow.
Morlaix Dowlas broad.
Ditto narrow.
Creas broad.
Ditto narrow.
Ditto of *Gascony*.
Coletas broad.
Ditto narrow.
Cotences fine.
Ditto ordinary.
Sail-Cloth.
Combs.
Haberdashery.
Gold and Silver Lace.
Silk Lace.
Fine Thread Lace, or Bone Lace.

From HAMBURGH.

Pipe-ſtaves large.
   Ditto ſmall.
Ordinary Boards of *Norway*.
Ditto of *Sweden*.
Great Planks.
Platillas, or, blue Paper Sleſies.
Bocadillos.
Eſtopillas.
Capadereys.
Creſuelas white.
Ditto brown.
Checker'd Linnen in Rolls.
Veſtualias.
Bed-Tickin fine.
Ditto ordinary.
Napkins and Table Linnen.
Eſterlines.
Fuſtians double.
Ditto ſingle.
Latten Wire.
Starch.
Powder-blew.
Gilt Leather.
Leaf-gold.
Pins.
Braſs Wire.
Braſs Weights and Scales.
Braſs Kettles and Pans.
Yellow Wax of *Dantzick*.
Barbary Wax whiten'd.
Caſes of Bottles.

From ITALY.

Ribbons of all sorts.
Hair Chamlots.
Silks flower'd with Gold and Silver.
Naples Silk.
Shags.
Velvets, one and half, and two Piles.
Grograms of *Messina*.
Men, Women, and Childrens Stockings of *Naples*.
Genoa Paper.
Hose of course Silk, call'd *Capullo*.
Mohair Stuffs from *Smyrna*.
Genoa Thread.
Ditto of *Salo*.
Iron Wares of *Genoa*.
Rice of *Milan*.
Hard Soap of *Genoa*.
Hoops.
Wheat of *Sardinia*.
Alom.
Brimstone.
Annifeed and other Seeds.

From PORTUGAL.

Musk in Cods.
Amber-grease.
Civet.
Fine Thread.

From SPAIN.

Taffaties of *Granada* double.
Ditto single.
Ditto of *Jaen*.
Ditto of *Antequera*.

Black

Black Silk of *Granada*.
Ditto colour'd.
Sattins flower'd.
Ditto plain.
Toledo silk Hose.
Mix'd Serges of *Ampudia*.
Saffron of *Villa Alva*.
Hard Soap of *Alicant*.
Almonds of *Alicant*.
Ditto of *Valencia*.
Wooll.
Wheat.
Barley.
Xeres Wines.
Tent Wine of *St. Lucar*.
Oil of *Sevil*.
Figs.
Raisins of *Arcos* in Barrels.
Ditto in Frail.
Salt.
White Wax.
Iron Ware from *Biscay*.
Steel.

A List of Commodities brought from the Spanish West-Indies *into* Europe.

Pearls.
Emerauds.
Amethists.
Virgin Silver.
Ditto in Pigs.
Ditto in Pieces of Eight.
Virgin Gold.

Ditto

The Introduction. xvii

Ditto in Doblones.
Cochinilla of several sorts.
Grana Silvestre, or, Wild Scarlet.
Ditto of *Campeche.*
Indigo.
Anatto.
Logwood.
Brasilette.
Nicaragua Wood.
Fustick.
Lignum Vitæ.
Sugars.
Ginger.
Cacao.
Bainillas.
Cotton.
Red Wooll.
Tobacco in Roll.
Ditto in Snuff.
Hides raw.
Ditto tann'd.
Ambergrease gray.
Ditto black.
Bezoar.
Balsam of *Peru.*
Ditto of *Tolu.*
Cortex Peruviamus, or, *Jesuit*'s Bark.
Jallap,
Mechoacan.
Sarsaparilla.
Sassafras.
Tamarinds.
Cassia.

The Introduction.

In the firſt Volume there is Mention made of one *Alexander Selkirk*, (ſo commonly call'd, but his right Name is *Selcrag*) who being left on the Iſland *John Fernandes*, continu'd there four Years and four Months, without any human Society. That ſhort Hint rais'd the Curioſity of ſome Perſons to expect a more particular Relation of his Manner of living in that tedious Solitude. We are naturally fond of Novelty, and this Propenſion inclines us to look for ſomething very extraordinary in any Accident that happens out of the common Courſe. To hear of a Man's living ſo long alone in a deſert Iſland, ſeems to ſome very ſurprizing, and they preſently conclude he may afford a very agreeable Relation of his Life, when in Reality it is the moſt barren Subject that Nature can afford. Even this ſolitary Life is not ſo amazing; we have in the aforeſaid firſt Volume mention'd two other Perſons, who at ſeveral Times continu'd long on the ſame Iſland, and without thoſe Conveniencies this Man we here ſpeak of was furniſh'd with; and yet it was never thought worth while to give any particular Account of their Behaviour there. Not to confine our ſelves to them, we have the written Lives of ancient Anchorites, who ſpent many Years in the Deſerts of *Thebaida* in *Egypt*, without
ſeeing

seeing any human Creature. The Lives of those holy Men, are little read or regarded, because they entertain us with nothing but a continu'd Course of Austerities and Devotion. From this Man something of another Nature is expected, his Piety is not likely to disgust us. What then can it be that flatters our Curiosity? Is he a natural Philosopher, who, by such an undisturb'd Retirement, could make any surprizing Discoveries? Nothing less, we have a downright Sailor, whose only Study was how to support himself, during his Confinement, and all his Conversation with Goats. It would be no difficult Matter to embellish a Narrative with many Romantick Incidents, to please the unthinking Part of Mankind, who swallow every Thing an artful Writer thinks fit to impose upon their Credulity, without any Regard to Truth or Probability. The judicious are not taken with such Trifles; their End in Reading, is Information; and they easily distinguish between Reality and Fiction. We shall therefore give the Reader as much as may satisfy a reasonable Curiosity, concerning this Man, without deviating into Invention.

The *Duke* and *Dutchess* Frigats coming up to the Island of *John Fernandes*, as was said in the first Volume, there appear'd

The Introduction.

pear'd on the Shore a Man waving a white Flag, which was the Stranger, becaufe the Ifland is known not to be inhabited. Some Officers went in the *Duke*'s Boat, and coming near the Shore, heard him fpeak to them in *Englifh*. They call'd to him to fhew them a good Place to come to an Anchor, and to land; he gave them Directions, and then ran along the Shore in Sight of the Boat, fo fwiftly, that the native Goats could not have out-ftripp'd him. When come to the Place, he faluted the new Comers with much Joy, being fatisfy'd they were *Englifh*, and they in Return invited him Aboard; he firft enquir'd whether a certain Officer that he knew was Aboard; and hearing that he was, would rather have chofen to remain in his Solitude, than come away with him, 'till inform'd that he did not command. Then the Officers that were in the Boat went afhore, whom he invited to his Habitation. The Way to it being very much hid and uncooth, only Capt. *Fry* bore him Company; and having with much Difficulty climb'd up and crept down many Rocks, came at laft into a pleafant Spot of Ground full of Grafs, and furnifh'd with Trees, where he faw two fmall Huts, indifferently built, the one being the Lodging Room, and the other the Kitchin. The Furniture, it may well be fuppos'd, was not extraordinary,

but

The Introduction. xxi

but confifted of every Thing that Defert could afford. There was in the Kitchin a Pot or Kettle to boil Meat, which that Inhabitant had carry'd afhore from his Ship; the Spit was his own handy Work, of fuch Wood as grew on the Ifland. His Bed rais'd from the Ground, on a Bed-ftead of his contriving, confifted of Goats Skins, the reft fuitable to the Habitation. About it was a Parcel of Goats he had bred up tame, having taken them young, which ferv'd to fupply him upon Occafion when he fail'd of any wild. He had provided fome of his Goat's Flefh to entertain his Guefts, which after their long Run at Sea, from the Ifland *Grande*, round Cape *Horn*, was no fmall Dainty. They had much Difficulty to perfwade him to venture himfelf Aboard, fo great was the Averfion he had conceiv'd againft the Officer aforefaid; yet, upon Promife of being reftor'd to his former Dwelling, if not fatisfy'd, he at length comply'd, and found fuch Enternent, as made him no longer fond of his folitary Retreat. The whole Account he could there give of his Manner of living on the Ifland, was in Effect, That having taken with him from aboard the Ship the *Cinque Port's* Galley, to which he belong'd, all the Neceffaries he could get towards providing for a Subfiftance in that abandon'd Place, he at firft had

been

been oblig'd to feed upon *Seals*, and such other Fish as he could take along the Shore ; which proving course Dyet, put him upon a Necessity of looking out for some Variety. There were Goats enough, but the Difficulty, was, how to catch them, among the Rocks and Mountains ; where, tho' shot, they would sometimes make their Escape into such Places where they could scarce be found. But Hunger is irresistible, which put him upon trying all Expedients for the Support of Nature. He us'd himself to running, and scrambling among the Rocks, 'till some of the tender Kids became a Prey to him, and by long Practice, at length improv'd so much, that the most nimble Goats could not escape him in their native Soil. He knew all the by Ways and Paths on the Mountains, could trip from one Crag to another, and let himself down the dreadful Precipices. Being arriv'd to this Perfection, his Life began to grow easier, as having Fish and Flesh for his Table. There still seem'd to be a Want of Bread, but Nature and the *Spaniards* had in some Measure supply'd that Defect, Nature by furnishing the Cabbage-Tree, describ'd Vol. 1. pag. 116. and represented there Plate 4. Numb. 8. the *Spaniards*, who first inhabited the Island, by leaving there the Seeds of Turnips, and several other Roots, which have since remain'd in the Ground.

The Introduction. xxiii

Ground. Such were the Provisions he had to feast on; for Conveniency of dressing, it has been said, he built a Kitchin, had a Pot or Kettle from Aboard, and the Trees, whereof there is Plenty and Variety, supply'd him with Spits, and Store of Fewel, having an Ax and some other Tools for that Purpose: The greatest Disaster he met with during his lonely Abode there, was, in hunting, when being once in eager Pursuit of a Goat, he dropt down from a Precipice, with such Violence, that he lay for a considerable Time as dead, and afterwards coming to himself, it was long before he could recover his Habitation. At last, no Help being to be expected, he crept Home, and there by Degrees recover'd of his Bruises, without the Assistance of Doctor, Surgeon, or Apothecary. I had forgot, in describing his Dyet, to speak of Drink; but that was the only Thing he could not want, the Island being sufficiently furnish'd with fresh Water, to satisfy his Thirst, without being tempted to Debauch. Some few *Spanish* Ships happen'd to touch there, during his Stay; but he had resolv'd rather to converse with his Goats, than be beholding to that Nation for his Deliverance from that Prison; and therefore, when their Ships appear'd, he generally kept close in his Apartment, which, as has been said, was so conceal'd, as not easily

The Introduction.

sily to be discover'd. However, being desirous to know what People they were who arriv'd in his Dominions, he kept not himself always so reserv'd, but that he was once 'spy'd and pursu'd, and some Shot made at him. His Activity then stood him in stead; for he out-stripp'd them all, and easily disappointed their Search; and the Prize being so inconsiderable, it is likely they thought it not worth while to be at any great Trouble to find it. Having little to divert his Thoughts, he had kept an exact Account of the Day of the Month and Week, all the Time of his Confinement, and told it to several of our Officers, when they first came to him on the Island. To conclude, he came away with us, and arriv'd safe in *England*, where he has freely imparted thus much, as he had done before Aboard, to all that have had the Curiosity to converse with him.

This may suffice as to him, being the whole material Truth, and sufficient on such an Account; and with it we will put a Period to this Introduction, to proceed with the Voyage where we left off.

A VOYAGE TO THE South Sea, &c.

VOL. II.

CHAP. I.

Departure from California; *the long Run a-cross the* South *Sea, to the Iſlands* Ladrones; *Arrival at* Guam, *one of that Number; courteous Entertainment there by the* Spaniards; *Letters and Certificates on both Sides; Variation in thoſe Parts,* &c.

THE great and wealthy Countries lying along the *South* Sea, being the Object on which the Eyes and Thoughts of all Men are at preſent fix'd, leſs could not well have been ſaid of them, than has been done in the firſt Volume

B   of

A Voyage *to the* South Sea, *and* of this Voyage. Our Ships ran along that Coast, to make such Attempts as were proportion'd to our Strength; which being but small, we had not the Opportunity of performing many Actions, that might of themselves have render'd this Work both useful and entertaining, there is a farther View in what may be perform'd hereafter; and a bare Run at Sea, could not have answer'd those Ends, which every Man may propose to himself in the Perusal of this Voyage. We brought our Ships Home from *California*, after taking the *Acapulco* Prize, that the Relation might not appear altogether maim'd and imperfect; but, at the same Time, we promis'd a more ample Account of their long Navigation in this second Volume. We had the greatest Part of the Globe to sail round, when we departed *California*; and so great a Tract will well deserve to be seen more at large, than could possibly be done before. This shall be the Subject of the first Part of this second Volume, which shall conclude, as was promis'd in the first, with a more ample Description of all the Coasts of the *South* Sea, with the Bearings of all the most noted Lands, and all curious Observations as to Winds, Currents, &c. taken from the *Spanish* Manuscripts. I will now proceed to the Journal.

Departure from California. Tuesday, *January* 10. 1709-10. The Ship *Marquis*, which I commanded, with the *Duke*, *Dutchess*, and *Manila* or *Acapulco* Prize, now call'd the *Batchelor* Frigat, being at *Puerto Seguro*, in the Island of *California*, in the Latitude of 23 Deg. 10 Min. *North*, ready to sail, Capt. *Rogers* put aboard the Prize about 30 Men, Capt. *Courtney* 25, and my self 12; which, with about 30 *Lascars* or *Indians*, and Blacks,

Blacks, made 100 Men. The Breeze beginning to come off from the Shore at nine in the Evening, we all unmoor'd, and at twelve weigh'd, and ran out with a small Gale at *N.W.*

Wednesday, January 11. had little Wind in the Morning, and sometimes calm, and Capt. *Rogers* sent his Boat aboard all the Ships, with the following Letter.

 Capt. Courtney, *Capt.*   *At Sea,* Jan. 11.
  Cook, *Mr.* Fry, *and*   1709-10.
  Mr. Stretton,

G*Entlemen, the Prize sailing so very heavy, it*   Short Al-
behoves *us to provide for a long Passage: Our*   lowance.
Allowance *of Flower now being two Pounds for five Men a-Day, is little; but having no Remedy, it must be less, and we ought to allow but one Pound and a half of Flower or Bread for five of our Men, and the same for six others. I do not doubt you'll agree with me, and have order'd that Allowance to begin this Day. I am your Friend to serve you,*

      Woodes Rogers.

This Day at Noon, Cape St. *Lucas* bore *North*, distant five Leagues. The Run from *California* being long, in an open Sea, and most before the Wind, there is not so much remarkable in it, as in other Passages of less Length; I have therefore here plac'd the following Journal-Table of the Voyage to the Island *Guam*.

This Table shewing our Course, the Winds, Latitudes, and other Particulars of that Nature, I shall proceed to what is otherwise material during the said Run to, and our Stay at the Island *Guam*, one of the *Ladrones*.

No Variation.

Sunday, January 15. 1709-10. I obferv'd that having had pleafant fmall Breezes of Wind at N. E. and fine Weather fince the 13th, and fteer'd S W. by S. yet made but a S. S. W. Courfe, it cannot be fuppos'd that the Variation fhould occafion the Miftake, becaufe that is there fo fmall as fcarce to be worth taking Notice of; and therefore it muft be of Neceffity caus'd by the Current. For Latitudes, &c. I refer to the Table.

Tuefday 17. faw feveral Sea-Fowls, which made me judge we were near fome Iflands. Thurfday 19. faw a Multitude of Craw-Fifh fwimming by us, and feveral Lumps, with Barnacles on them, which I did fuppofe to to be Amber-Greafe; but, for want of a Boat, could not take up any. Saturday 21. faw feveral Shoals of Flying-Fifh. Monday 23. our Ship proving very leaky, and no Poffibility of coming to ftop it, becaufe in or near the Stern, was forc'd to ftitch a Sail with Oakham, hang Weights to each Clew, and lower it down athwart the Cutwater, believing the Leak might fuck in fome of it. Wednefday 25. perceiving our Bonnet had done the Leak but little Good, ran a-head the other Ships and brought to; but the Sea ran fo high, that the Leak could not be ftopp'd; and continuing bad, I got down our Top-Gallant Yards to eafe the Ship. Friday 27. having brought the Ship more by the Stern, found fhe was not fo leaky as before.

Thurfday, February 2. 1709-10. a hard Gale of Wind at E. N. E. and a great Sea following us, with feveral Squals of Rain and Wind, which continu'd 'till the 4th, when we had moderate Breezes. Wednefday 8 ran a-head again to ftop the Leak; but there being a great
Sea,

Sea, could not come at it. *Friday* 10. being in 13 Deg. 45 Min. *North* Latitude, and 47 Deg. 40 Min. Longitude *West* from Cape St. *Lucas*, obferv'd by a good Amplitude, that we had half a Point Variation *Eafterly*

Tuefday, February 14. 1709-10 I reckon'd we were 3 Deg. 22 Min. fhort of 180 Degrees of Longitude from the Meridian of *London*, *Weft*, which would make half way round the Globe. About this Time had feveral Men ill, and fome dropp'd down at the Pump and Helm; which, I fuppofe, was occafion'd by the Badnefs and Shortnefs of Provifions, therefore began to allow them more. *Monday* 20. being in 13 Deg. 28 Min. Latitude *North*, and 69 Deg. 1 Min. Longitude, found above half a Degree Variation *Eafterly*. More Men fell fick. *Wednefday* 22. in 13 Deg. 8 Min. Latitude, and 72 Deg. 36 Min. Longitude from Cape St. *Lucas*, found about feven Degrees *Eafterly* Variation. *Friday* 24. ftill in 13 Deg. 8 Min. *North* Latitude, and 75 Deg. 22 Min. Longitude, nine Degrees *Eafterly* Variation. *Sunday* 26. ftopp'd one of our Leaks.

Wednefday, March 8. 1709 10. the Clouds fettled to the *Weftward*, we faw feveral Sea-Fowls, which made me conclude we fhould foon fee the Iflands *Ladrones*. *Friday* 10. at three in the Afternoon, made the Ifland *Sarpana*, one of the *Ladrones*, bearing *W. N W*. diftant 12 Leagues, and at Six the Ifland *Guam* bearing *W*. by *S*. diftant 12 Leagues. Lay by moft Part of the Night; and *Saturday* 11. towards Morning, made Sail, the Wind at *E. N. E.* a frefh Gale, fteer'd *S. S. W*. and *S. W*. At Noon, the *South* Part of the Ifland *Guam* bore *W*. by *S*. diftant three Leagues, and the *Northermoft* Part *N*. by *E*. diftant 8 Leagues.

Half round the Globe.

Arrive at the Iflands Ladrones.

Latitude *per* Eſtimation this Noon 13 Deg. 30 Min. Longitude from Cape St. *Lucas* 100 Deg. 19 Min. we gave the low Point at the *S.W.* a Berth, by Reaſon of the Shoals, then hal'd upon a Wind, and made ſeveral Trips. At Six in the Evening came to an Anchor in Port *Umatta* of the Iſland *Guam*, or *Guana*, in 15 Fathom Water, at about a Mile Diſtance from the Houſes, which bore *E*. by *N*. another Part of the Iſland bearing *South*, diſtant two Miles, and the Rock to the *Northward*, *N.E.* by *N*. from which runs a Ridge of Rocks to the Iſland, and a Shoal to the *Eaſtward*. We went in with *French* and *Spaniſh* Colours, that they might not ſuſpect us to be Enemies. Several of their Boats, which they call *Paraos*, came about our Ships, but none would venture aboard, 'till, being off the Anchoring-Place, one was ſent by the *Spaniſh* Governor, deſiring to know who we were, what we wanted, and to ſhew us where to anchor. We preſently ſent away Mr. *White*, our Interpreter, and one *Murphy*, an *Iriſhman*, taken in the *Batchelor*, with a Letter to the Governor, and detain'd a *Spaniard*, as Hoſtage, 'till their Return. He ſent a very obliging Anſwer, promiſing to ſupply us with what the Iſland afforded, and ſending ſome Refreſhments at the ſame Time. Our Letter to him, and his Anſwer, were as follows.

The Letter to the Governor.

SIR,

WE *being Servants to her Majeſty of* Great Britain, *and oblig'd to ſtop at theſe Iſlands in our Paſſage to the* Eaſt-Indies, *will not moleſt the Settlements, provided you deal friendly by* by us, and *shall pay for all Provisions and Conveniencies, either in Money, or such Necessaries as you want. But if, after this civil Request, you deny us, and act not like a Man of Honour, you may expect such military Treatment, as we are with Ease able to give. This we have thought fit to give under our Hands, recommending to you our Friendship and kind Treatment, which we hope you'll esteem, and assure your self we shall then be with the strictest Honour,*

To the Honourable   *Your assur'd Friends, and*
the Governor of
the Island *Guam*.   Humble Servants,
March the 11th.
1709-10.

     Tho. Dover, | Woodes Rogers,
     Steph. Courtney, | Edw. Cooke.

The Governor's Answer.

Gentlemen,

I *Receiv'd a courteous Letter from you, the Bearer whereof acquainted me with your great Wants, and that you requested Refreshments, Wood, and Water; I answer you with the same Civility, and offer all I possibly can perform; but we have had a violent Distemper here, and bury'd abundance of our People. Tho' you are our Enemies, upon your paying for what you have, as you write in yours, I have order'd all under my Command to offer no Injury to any of yours, and desire you will do the same, permitting to pass to and fro without Molestation. Capt.* Don Antonio Gutieres *is my Friend, I have some Business with him, and desire you will let him come to me, and Capt.* John Antony Peftana *will remain in his stead. I also desire you will let me have all the* Spanish *Prisoners.*

I have

I have order'd Capt. J. R. to supply you with what we can.

To the four Captains,   Yours,
viz. *Woodes Rogers,*
Steph. Courtney, Tho.
Dover, and *E. Cooke.*
           Don Juan Antonio Pimentel.

Sunday, March 12. 1709-10. our Pinnaces went a-shore to the Watering-Place with some of our Officers, where they were civilly treated by a *Spanish* Captain, and other Officers at the Port of *Umatta.* The Latitude and Longitude of this Place, is laid down wrong in most of our Books and Charts, except Capt. *Haley*'s Variation-Chart, which comes very near the Matter, as well in this Run, as in our others; but none of our Authors that I have read, take any Notice of the Variation in these Parts. and we find it half a Point *Easterly* at this Island; and in our Passage to it, we had sometimes 12 Degrees. The Reason of this I take to be the Unevenness of the Globe of the Earth, and its unequal Mixture of much Matter differing in it self as to the magnetical Quality; as having large and stony Mountains, spacious Valleys, deep Seas, long Continents, high Promontories, with mighty scatter'd Rocks of Load-stone, Iron Mines, and other magnetical Substance.

We continu'd at *Guam* 'till *Tuesday, March* 21. 1709-10. which Time was spent in fitting, wooding, watering, and carrying off Provisions and Refreshments. The *Spanish* Gentlemen there treated us with all imaginable Civility, and supply'd us with the following Quantity of Provisions, asking nothing
for

for them; but we apprais'd the said Provisions, and return'd the full Value in Goods they stood in need of, to their extraordinary Satisfaction.

Provisions taken in at the Island Guam, and their Value.

	Dollars.
4000 Coco Nuts, at 10 Dol. *per* 1000,	0040
100 Baskets of *Maiz*, or *Indian* Wheat, at 1 Dol. 4 R. *per* 2 Baskets,	0075
60 Bags of Rice and Paddy, at 2 Dol. *per* Bag,	0120
250 Baskets of Yams and Patatas, at 4 R. *per* Basket,	0125
360 Fowls small and great, at 2 Roy.	0090
220 Hogs small and great, at 4 Dol.	0880
51 Bullocks, at 14 Dol.	0714
8 Cows with Calves, at 18 Dol.	0144
300 of Eggs, at 2 Dol. *per* 100,	0006
	2194

Liquors, &c.

3 Cases of Brandy, each containing 15 Bottles, at 15 Dol. *per* Case,	0045
4 Jars of Coco Wine, at 15 Dol. *per* Jar,	0060
3 Jars of *Nipa* Wine, at 10 Dol. *per* Jar,	0030
11 Jars of Bread, at 3 Dol. *per* Jar,	0033
3 Jars of Sugar, at 5 Dol. *per* Jar,	0015
2 Jars of Vinegar, at 5 Dol. *per* Jar,	0010
	0193
	2194
Total,	2387

Value

Value of Returns made for the above Provisions, &c.

	Dollars.
2 black Women, at	0225
2 black Men, at	0225
40 Pieces of good Bays, at 25 Dol. per Piece,	1000
12 Pieces of damnify'd Bays, at 10 Dol. per Piece,	0120
6 Pieces of scarlet Shaloon, a little damag'd,	0180
20 Match-Locks, at 12 Dol. each,	0240
1 Box of Medals,	0120
1 Box of Relicks, Pictures, &c.	0157
6 Boxes of Nails, at 20 Dol. per Box,	0120
Total,	2387

2 black Boys, &c. a Present to the Governor.

The Governor, being lame, could not come Aboard us, but several of our Officers went to see him, whom he receiv'd and treated very civilly, we sending by them as a Present two black Boys in Liveries, and some other Things; for he liv'd not at the Port, but farther up to the *Northward*. None of our chief Commanders went to see him, but the Gentleman who is next to him in Command, came several Times, with other Officers, Aboard our Ships, and we went daily ashore to them, entertaining one another. *Monday* the 20th of *March*, all the *Spanish* Gentlemen, and most of our Officers in chief, were Aboard together, to conclude all Affairs among us, where it was agreed to give Certificates on both

Round the World.

both Sides, of the good Usage each Party had receiv'd from the other, and were as follows:

     Guam, March 21. 1709-10.

WE the Commanders and chief Officers of four British private Ships of War and Prizes, do hereby acknowledge, that arriving at the Island of Guam, and in Want of Refreshments, we met with a kind and generous Reception from the Honourable Don Juan Antonio Pimentel, Governor and Captain General of the Marian Islands, and Capt. Don Juan Antoino Pretana, and other Officers and Gentlemen of the said Island, and were plentifully supply'd, in a shorter Time, and better Manner, than we could have expected. During our Stay here, we liv'd in a very Friendly Manner, and at our Departure, made such Presents to the said Gentlemen, in Return for the Necessaries they furnish'd us with, that they express'd themselves fully satisfy'd and contented therewith under their Hands, as on our Part we do the same,

William Dampier,	Thomas Dover,
Robert Fry,	Woodes Rogers,
William Stretton,	Stephen Courtney,
	Edward Cooke.

From the Lieutenant General *Don Juan Antonio Pimentel*, Governor and Captain General of the *Marian* Islands, &c.

    In his Name,

CAPT. Don Juan Antonio Prettana *informs, That four* English *Ships arriving at these Islands, whose Captains, are,* Woodes Rogers, Stephen Courtney, Thomas Dover, *and* Edward Cooke, *who came hither for Want of Provisions, they requested with much Courtesy, that we would*

A Voyage *to the* South Sea, *and* would spare them as much as we could, and they pay'd very liberally for it, more than double the Value, and treated all the Captains of this Island, with which they are well satisfy'd, and give it under their Hands,.

> Don Juan Antonio Prettana,
> Don Sebastian Luis Romez,
> Don Nicholas de la Vega,
> Don Juan Nunez.

The Surgeons also belonging to the Ships, sign'd the following Certificate.

WE *the Surgeons of the* Duke, Dutchess, Marquis, *and* Batchelor, *having view'd Senor* Antonio Gomez Figueroa, *a* Spaniard, *it is our Opinion, that he cannot live long at Sea, and therefore acquiesce with the Commanders to let him go ashore at the Island of* Guam, *to recover his Health, there being no Probability of carrying him to Great Britain. Witness our Hands, this* 21*st Day of* March 1709-10.

> James Wasse,
> John Barry,
> John Ballet,
> Charles May.

The Day before the signing of the above Certificates, a Committee was held on Board the *Marquis*, the Result whereof was as follows.

IT *is agreed, that we shall steer from hence a* W. *by* S. *Course, to go clear of some Islands that lie in our Way; and then we think it proper to steer a direct Course for the* S. E. *Part of* Mindanao, *and from thence the clearest Way to* Ternate.

Round the World.

nate. *It is also farther agreed, that Capt.* Rogers *shall deliver to Morrow Morning unto Capt.* Courtney, *one Chest of Plate and Money, to be put on board the* Dutchess.

	W. Stretton,	T. Dover, *Presid.*
		Woodes Rogers,
Joh. Ballet,	Cha. Pope,	Steph. Courtney,
W. Dampier,	T. Glendal,	Edward Cooke,
	J. Connely,	Robert Fry.

The Prisoners taken Aboard the *Acapulco*, or *Manila* Ship, except such as were necessary to condemn the said Prize, were, according to our Promise made to the Governor, set ashore.

CHAP. II.

A brief Account of the Marian *Islands, commonly call'd* Ladrones; *Description of the Island* Guam, *or* Iguana; *the Islands of* Solomon; Paraos, *a Sort of Boats; the* Rima, Ducdu, Areca, *and* Pine-Apple *Fruits,* &c.

THE Island *Guam* is one of those Ladrones most generally known by the Islands. Name of *Islas de los Ladrones,* or Islands of Thieves, from the natural Inclination of the Natives to stealing. The famous *Magellan,* who first sail'd through the Streight of his Name, into the *South* Sea, was the first Discoverer of these

14   *A* VOYAGE *to the* South Sea, *and*

these Islands, and gave them this Name of *Ladrones,* because the Natives coming Aboard his Ships, would snatch up every Thing of Iron they could lay hold on, and leap over Board; which Practice they afterwards continu'd with other Ships passing by, and by that Means confirm'd the Denomination. The same *Magellan* call'd them also *Islas de las Velas,* or Islands of Sails; from the great Number of *Paraos,* or Boats resorting to his Ships, which had three-corner'd Sails, made of Mats. The *Spaniards,* who are the only constant Traders in those Parts, having found it convenient to settle on some of them, for supplying of their Ships with Provisions and Refreshments, have rejected both those Names, and given that of the *Marian* Islands, that is, Islands of St. *Mary.* Their Distance from *New Spain* is generally reckon'd between 2300 and 2400 Leagues; and, by our Reckoning, from Cape St. *Lucas* in *California,* it is about 2000 Leagues. The constant Trade-Winds which reign between the Tropicks, are the Occasion of rendering this long Run extraordinary easy; and it is generally performed in about 60 Days, some few over or under. As to their Number and Position, they will be best seen in the Charts and Maps of the Islands of *India.*

Guam Island.

The Island where our Ships now anchor'd, is call'd *Guam, Guana,* or *Iguana,* lying in 13 Deg. 30 Min. of *North* Latitude, and 100 Deg. 20 Min. Longitude from Cape St. *Lucas* in *California,* and bears S. S. W. from *Sarpana,* another of the *Ladrones,* distant 8 Leagues. The S. W. Part of it is high, but the N. E. is low Land. The Valleys are very pleasant, having curious open Plains, Rivulets, and in some

some Places fine Groves of Trees. The Length of the Island from *N.E.* to *S.W.* is about 10 Leagues, the Breadth six. The Land on the Hills is red, and in the Vales a good fat black Soil; which, if manur'd, would certainly produce any Thing that is necessary for the Support of human Life. At present it affords Rice, Plantans, Bananas, Yams, Patatas, Rima, or Bread Fruit, Ducdu, Pine-Apples, Coco Nuts, Areca Nuts, Maiz, or *Indian* Wheat, Indigo, Jacas, Oranges, large Lemons, Limes, Guavas, Papas, Chaddocks, Water and Mush Melons, small *Garvanzos*, that is, *Spanish* Pease, Capers in Abundance, and Pompions. For Beasts, they have Black Cattel, Horses, Swine, Sheep, and Deer; and of Fowl, Hens, Pigeons, Curlieus, a Sort like our Black-Birds, and Thrushes. There is no great Variety of Fishes, the chief Sorts being Grampusses, Thrashers, Flying Fish, some Eels, and small Fish like Old Wives, Crawfish, and a small Sort like our Gudgeons, in the fresh Water Rivers and Brooks; and among the Rocks, by the Sea-side, large Scollops. Venemous Creatures there are none, except the *Centipes*. *Product.*

Beasts, Birds, and Fishes.

At the Port of *Umatta*, and another Place where the Governor resides, there are about 100 *Spanish* Soldiers, commanded by the Captain-General of these *Marian* Islands. In this Island there are two Churches, and three *Jesuits*, who instruct the *Indians* in the Christian Faith. The King of *Spain* is at the Charge of maintaining them, and the Garrisons here and at *Sarpana*, only these two Islands being at present inhabited by the *Spaniards*, as they have been for these 40 Years past, to supply their Ships trading from *New Spain* *Garrison and Churches.*

Spain, to the *Philippine Iflands* with, Refreshments. From hence runs a Ridge of Iflands at some Diftance from each other to the *Northward*, to above 20 Degrees of Latitude, and to the *Southward* there are others scatter'd about almoft as far as the Coaft of *New Guinea*.

I have here given a View of the Ifland *Guam*, or *Iguana*, as taken at Port *Umatta*. Plate 1.

Iflands of Solomon.

The *Spaniards* at *Guam*, or *Iguana*, inform'd us, that a Ship of theirs failing formerly from *Manila* for *New Spain*, difcover'd a Parcel of Iflands very pleafant, and abounding in Gold, Amber-Greafe, and other valuable Commodities, and gave them the Name of the Iflands of *Solomon*. They fay feveral Ships have been fince fent out in queft of thofe Iflands, but could never find them; and fom *Paraos*, which have ventur'd upon the fame Difcovery, not knowing how to fteer, when out of Sight of Land, have never more been heard of. Thefe Iflands fome place in 15 Deg. 20 Min. of *North* Latitude, and 300 Leagues to the *Eaftward* of the *Ladrones*. Others will have them to be in 14 Degrees *South*. The aforefaid Ship having been drove by Strefs of Weather upon an Ifland, it appear'd that the Agitation of the Veffel had remov'd all the Earth from about the Hearth of the Furnace, which was fupply'd with fome taken from the faid Ifland. When this Ship arriv'd at *Acapulco*, removing that Earth, they found under it a Mafs of Gold, which the violent Heat of the Furnace had melted, and feparated from the Earth. The Commander furpriz'd at this unexpected Accident, acquainted the Vice-Roy of *Mexico*, and he the King, who fitted out fome Ships to find the faid Ifland. They

kept

Plate 1.

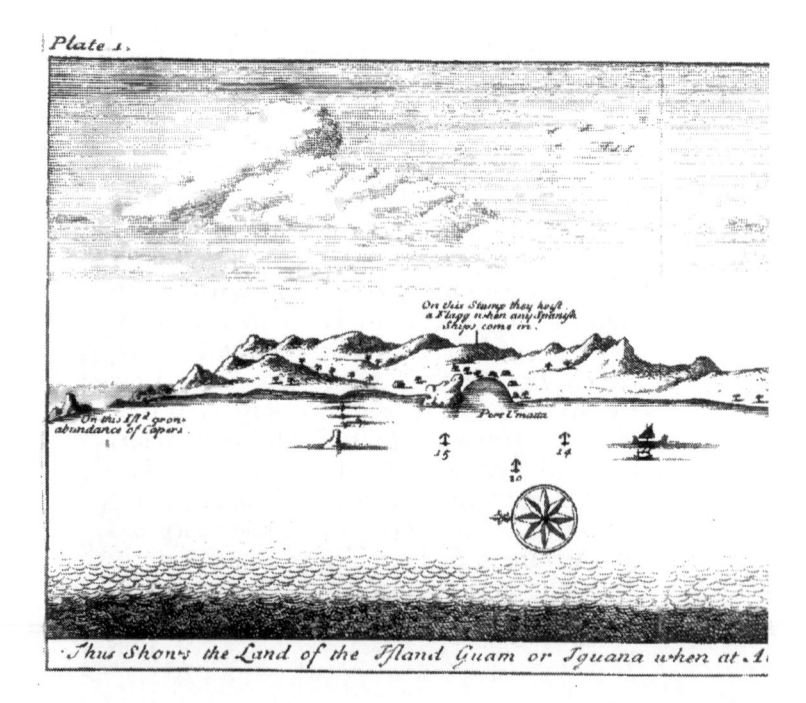

Thus Shows the Land of the Island Guam or Iguana when at A...

Spain, to the *Philippine Islands* with, Refreshments. From hence runs a Ridge of Islands

from the Earth. The Commander surpriz'd at this unexpected Accident, acquainted the Vice-Roy of *Mexico*, and he the King, who fitted out some Ships to find the said Island. They
kept

kept to the *Southward* of the Line, and could not find it, only one of those five Ships returning Home to *New Spain*. I am of their Opinion, who believe it must be to the *Northward* of the Line, because the Ship which is said first to have found it, being bound for *New Spain*, must of Necessity keep well to the *Northward*, else could not make the Passage by Reason of the Trade-Winds between the Tropicks.

This my Opinion is grounded, as above, on the suppos'd Discovery made by the Ship bound from *Manila* for *New Spain*; however, the Fragment we have of the Discovery made of the Islands of *Solomon*, seems to place them to the *Southward* of the Line, which was by Ships sent from *Peru*; and in it there is no Notice taken, where they mention Latitudes, of their passing over to the *Northward*. That Relation is so maim'd and imperfect, that it gives us little Light for finding of those Islands, which are there very advantageously represented; but since we have so little Knowledge of them, it will be proper to return to the *Ladrones*, or *Marian* Islands.

The Natives of them are of a dark Complexion, not so black as the *Indians* of *California*, but most of them the largest and best limb'd Men I ever saw, and some of them very hairy and strong. The Women are strait and tall; near about where the *Spaniards* reside, they have something to cover their Privities, but a League farther up, they go stark naked, both Men and Women. The savage Part of them are said to eat white Men, if they take them, and drink their Blood, devouring all they catch raw. Some have no peculiar Worship; but the most pay their Adoration

Natives of the Ladrones.

Vol. II.   C   to

18 *A* Voyage *to the* South Sea, *and*
to the Sun, the Moon, and several other Creatures, according to every Man's particular Fancy. One arm'd *Spaniard* will beat 40 of them. They are very dexterous at catching of Fish, and building and managing their flying *Paraos*, which I shall describe below.

Temperature.
We found the Weather very hot here, and yet the Island is counted very healthy, by Reason of the fresh Trade-Wind continually blowing; however, many of the peaceable *Indians* have dy'd of late of the Leprosy, and I saw several who had it when we were there; but all the White Men were clear.

There is good Anchoring here, in clean Ground, within less than a Mile off the Shore, in 10 or 12 Fathom Water, right off the Village of *Umatta*, the Wind generally blowing off the Shore. The little Mr. *Funnel* says of this Island, and Parts about it, is not to be regarded, being all contrary to what I have found, and perhaps only taken upon Hear-say. *Cowley*'s Voyage has as little of Truth; for he makes it 14 Leagues long, and talks of 600 *Spaniards* in Garrison there; which is all false, as may be seen by what has been said above.

Umatta Port.
At *Umatta* there is a large House for the Governor, built after the *Spanish* Fashion, with Galleries about it, for Coolness. The Church is of Boards and Bamboes split, and cover'd with Palmito Leaves, as is the House for the Priest, and the Guard-House; besides which, there are several Pens for Cattel and Fowl, and many *Indian* Huts, all of Bamboe and Palm-Tree Leaves. About these Houses grow the several Sorts of Fruit above-mention'd.

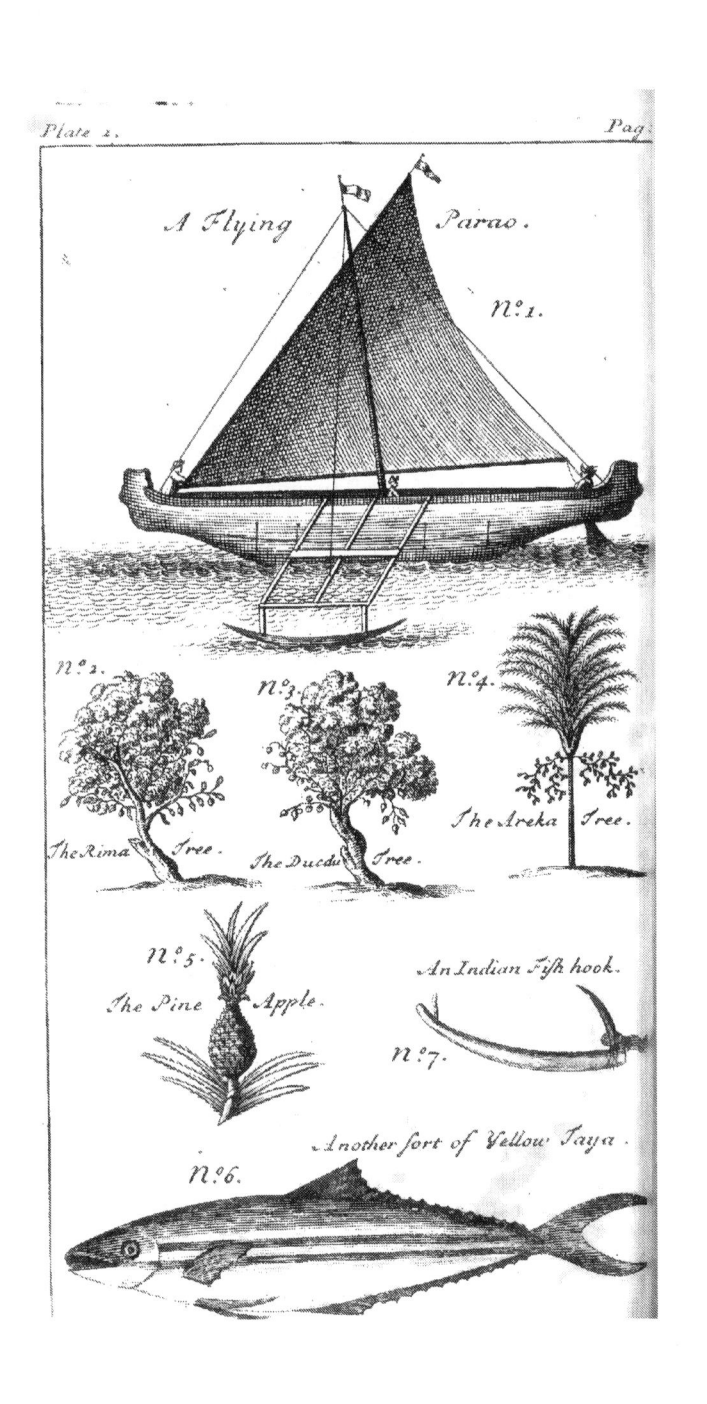

Plate 1. Pag.

A Flying Parao. Nº 1.

Nº 2. The Rima Tree.
Nº 3. The Ducdu Tree.
Nº 4. The Areka Tree.
Nº 5. The Pine Apple.
An Indian Fish hook. Nº 7.
Another sort of Yellow Jaya. Nº 6.

A Ship comes hither once a-Year from *Manila*, with Necessaries for the *Spaniards* inhabiting the two Islands. I saw no Fortification at this Place; but it is likely, and we are told by others, that there is a small Castle up the Country, where the Governor resides, to curb the Natives.

The flying *Paraos* of this Island, are very unaccountable, as well for their strange Make, as for their extraordinary swift sailing. They are made of two Trees hollow'd, like a Canoe, and sew'd together with strong fine Sinnet, made of the Threads of some of the Palm-Trees. When laden, they draw between a Foot and a half, and two Foot Water, being built sharp, and not above three Inches broad at the Bottom, and 20 Inches at the Top, and about 35 Foot in Length. Being so narrow, they have two Out-leakers, eight Foot distant, plac'd in the middle, always on the Weather-side, and about 12 Foot long, with a Log at the Ends of the Out-leakers, made fast with Stantions of two Foot long to the Log, which is about 15 Foot long, and made in the Shape of the Bottom of the *Parao*, which always swims in the Water, and keeps her steady; the Lee-side of the *Parao* being built near upon a Line, and the Weather-side rounding, for which Reason they are the best Boats in the World upon a Wind. They have but one Mast of about 20 Foot long, and a Sail made of a Mat, three corner'd, like that of a Settie, about 21 Foot deep at the Leech, with a Yard 25 Foot long, and a Boom of the same Length at the Foot of the Sail, with the Sheet made fast two Thirds out of the Boom, the Sail being lash'd both at the Yard and Boom, and the Boom

Paraos Boats.

lash'd

lafh'd at the Foot of the Yard. No Sort of Boats whatfoever can come near them for Swiftnefs in Sailing; for by Report of the *Spaniards* of this Ifland, they will run above 20 Leagues an Hour. *Don Juan Antonio Pretana* told us, he would lofe his Head if they did not perform it; and becaufe we thought it incredible, he affirm'd that one we had prefented us, and defign'd to bring for *England*, would fail 30 Leagues in an Hour. I could not believe it; but have feen them fail at a prodigious Rate. When they turn to Windward, and defign to ftand the other Way, they let go the Sheet, and fhift the Tack to the End the Sheet was at, placing the Tack, or End of the Yard, in a Notch cut in the Thaughts at each End of the *Parao* for that Purpofe; and that which was the Stern before, thus becomes the Head; either End going foremoft, there being no Difference in the Built of them, but the fame ftill remains, and always is the Weather-fide. Moft of thefe Boats are painted red above the Water, and black below. A Board, about eight Inches broad, is made faft on the Weather-fide, from End to End, to keep the Sea out; and for carrying of Goods or Paffengers, they lay Boards a-crofs the Out-leakers, about two Foot out from the Side of the *Parao*, where they place them hanging over the Water. There are generally three *Indian* Sailors in thefe Boats, one being always in the middle to lade out the Water, which comes in thro' the Seams, and over the Sides; the other two fit at each End of the Boat, to fteer with a Paddle in the Lee-Quarter, fhift the Sail, and hale aft the Sheet. The Maft ftands with a Fork at the End of it, upon the middle Piece of the Out-leaker, from

which

which comes a forked Pole, made faft four Foot up from the Step of the Maſt, to keep it from falling to Windward. The Maſt always hangs forward; when they ſhift the Sail, and ſtand the other Way, they eaſe one Stay, and hale the other forward for that Purpoſe. See it exactly repreſented, Plate 2. Numb. 1.

I ſhall only add three or four Sorts of Fruit I took moſt particular Notice of. The *Rima* is a great Tree, as big as the Walnut, with large Leaves, having five Indentures on each Side. The Flower is oblong, of a yellow Colour, and a Pith, or Down within. The Fruit is as big as a Man's Head, of a Date Colour when ripe, and a rough Outſide; which boil'd or bak'd, is us'd inſtead of Bread, and ſerves the Natives for ſix Months. Cut in Slices, and dry'd in the Sun, it eats like Biſket. The Leaves ſerve the Cattel for Provinder, as well as thoſe of the *Ducdu.* See it Plate 2. Numb. 2. *Rima, or Bread-Tree.*

The *Ducdu* is like the *Rima*, both in Tree and Fruit, bating that the latter is more oblong, and has about 14 or 15 Kernels in it, about as big as a Cheſnut, and taſte very like them when roaſted, being all that is eaten of this Fruit. The Leaves are not ſo much indented as thoſe of the *Rima*. You have the Figure of it Plate 2. Numb. 3. *Ducdu Tree.*

The *Areca* Tree is like the *Palm*, but ſlenderer, and not ſo high. It bears a Bunch or Cluſter of Fruit or Nuts, incloſ'd in a Caſe or Huſk, like the Coco-Nut, about as big as a Nutmeg, and not unlike it when cut. This Nut is chew'd with *Betele* Leaves and Lime all over *India*, making the Teeth black and the Lips red; but reputed by the Natives, and *Areca Tree.*

C 3 all

all that use it, an excellent Preservative against the Tooth-ach and Scurvy, most of them being free from rotten Teeth, tho' of a great Age. It is represented, Plate 2. Numb. 4.

Pine-Apple Fruit. The Pine-Apple grows on a Stalk, about two Foot above the Ground, from amidst a Parcel of Leaves, not unlike the *Sempervivum*, or our *House-leek*, only the Leaves are two or three Foot long. The Fruit is oblong, about a Span in Length, yellow within, with Knobs or Squares on the Outside; whence it has the Name of a Pine-Apple, because resembling those which grow on our Pine-Trees. When ripe, it is yellow and red, with a Tuft of Leaves on the Top. It's Taste partakes of the Sweet and Sowre, with a most delicious Flavour, extraordinary pleasant. Some eat it with Sugar and Water. It is reckon'd very wholesom, tho' of a very hot Nature, insomuch that they affirm a Knife left sticking in it a whole Day, loses its Temper; yet it has no hot biting Taste, and is esteem'd by all *Europeans* as a most excellent Fruit. See this, Plate 2. Numb. 5.

Yellow-Tail Fish. The only particular Sort of Fish I took Notice of here, is a Species of *Yellow-Tail*, about 12 Inches long, and three in Breadth, having a small long Head, with a large Mouth and Eye, a Feather Fin on his Back, which runs to his large forged Tail. His other Fins and Tail as in the Figure, Plate 2. Numb. 6. His Back of a dark Yellow, the Belly of a Silver Colour. On his Sides, from Head to Tail, has two Streaks of Blue, and three of Yellow, the Tail and Fins Yellow.

A Fish-hook. The Fish-hooks these People use, are made of a large Bone, with a small one fix'd in it, looks white, and when tow'd in the Water, the

Round the World.

the Dolphin takes Hold of it, and is caught. See it reprefented Plate 2. Numb. 7.

CHAP. III.

Sail from the Ifland Iguana; *fee fome fmall Iflands; an Account of Spouts;* Moratay *and* Gilolo *Iflands; of the Monfons; Signals; dreadful Weather;* Mindanao *and the* Philippine *Iflands; their Trade.*

Tuefday, March 21. 1709-10. in the Morning the *Duke* and *Dutchefs* fir'd each of them a Gun, as a Signal to unmoor. At Eight the *Dutchefs, Batchelor,* and *Marquis* weigh'd, and made eafy Sail. Soon after the *Duke* came off with a fmall Breeze of Wind at *N. E.* At Noon the Body of the Ifland *Iguana* bore *E. N. E.* diftant fix Leagues.

Thurfday 23. from our leaving the Ifland *Iguana,* to this Day, had moderate Gales of Wind at *N. N. E.* with fome Showers of Rain, and clofe hot Weather, and fteer'd away *W. S. W.* with half a Point Allowance for the Variation. Our Courfe to this Day Noon but *W.* by *S.* Diftance 210 Miles, *Wefting* 200, *Southing* 60, Latitude *per* Obfervation and Eftimation 12 Deg. 30 Min. *North,* Meridian Diftance from the Ifland *Iguana* 3 Deg. 30 Min. *Departure from Guam.*

Friday 24. at Night, by a good Amplitude, found we had ftill half a Point of *Eafterly* Variation, and believe we had a Current that fet us to the *Northward.* Courfe to *Saturday* 25

at

at Noon *W.* quarter *S. Sunday* 26. in the Morning, judging our selves to the *Westward* of the Islands *Saavedra,* &c. steer'd away *S. W.* by *S.* and at Night, by a good Amplitude, had five Degrees of *Easterly* Variation, Latitude 10 Deg. 45 Min. *North,* Meridian Distance from the Island *Iguana* 8 Deg. 35 Min. *West.* Since our leaving that Island, Capt: *Courtney,* in the *Dutchess,* kept a-head, with a Light by Night, and every Day made Sail a-head of the other Ships, to discover any Danger before Night, and in the Evening brought to. I kept between him and the Prize, bringing to when the *Dutchess* did, and putting out Lights for the Prize; and when she came in Sight, made Sail again towards the *Dutchess.*

Thursday, March 30. 1710. we had now passed several small scattering Islands, as the *Matalotes, Arrecifes,* &c. but saw none of them, which we must have done, had they been right laid down in our Charts; yet I suppose we could not be far from them, having seen several small Birds. This Day Capt. *Courtney* and I went Aboard the *Duke,* and there agreed to steer away half a Point more *Westerly,* being to the *Southward* of the Shoals which lie off the aforesaid Islands. We farther agreed, as follows:

Signals for Keeping Company, &c.

THE Dutchess *to be a-head, the* Duke *next, the* Marquis *third, and the* Batchelor *last, all at a convenient Distance.*

In case the Dutchess *saw any Danger in the Night, she was to show a Light over the Poop-Light, and fire a Gun, making an easy Sail from it, so that the rest might be near enough to have sufficient Warning. Each Ship to answer with two Lights, and fire a Gun.*

In case the head-most Ship found it most proper to lie by, she was to show another Light, and fire two Guns, the least Light to be kept at the Boltsprit-Head; but if to continue under Sail, only the first Signal to be made. When all Signals were answer'd, so as to satisfy each Ship, then to keep a single Light out all Night; and if the head-most Ship, or any other, found the first Soundings in the Night, she was to show three Lights of an equal Height on her Poop, or Bow; if less than 30 Fathom, to show three Lights, two equal, and one over them. The Dutchess generally had a Light at the Mizen-Peak, which was not design'd for a Signal, but to know the Ship.

Monday, *April* 3. 1710. These Days past the Weather extreamly hot, several Sharks and other Fish were about our Ships, some of which we took. Judg'd we had a strong Current setting us to the *Southward*, farther than could be expected, which I perceiv'd by an indifferent Observation this Day at Noon, having had none some Days, by Reason the Sun was near the *Zenith*; found now 5 Deg. 45 Min. Latitude *North*, Longitude from the Island *Iguana* 12 Deg. 18 Min. *West*.

Wednesday 5. Latitude *per* Observation and Estimation 3 Deg. 45 Min. *North*, Longitude from the Island *Iguana* 13 Deg. 8 Min. *West*; perceiv'd a strong *Northerly* Current, and at least five Degrees Variation.

Monday 10. split my Main Top-Sail, and was forc'd to bend another. At Two in the Afternoon made a pleasant small low Island, bearing *E. S. E.* distant about six Leagues, not laid down in any of our Charts. It is very low Land, and full of Trees; and I made it to lie in the Latitude of 2 Deg. 55 Min.

North,

26 *A* VOYAGE *to the* South Sea, *and*

North, and of Longitude from *Iguana* 14 Deg. 40 Min. *West*. At Night we spoke to one another, and agreed to steer away S.W. 'till Ten, and then to lie by all Night, there being Danger, if we should meet with any more such Islands in the Night; for we saw the Trees long before the Land.

Islands and Spouts. *Thursday* 13. at Two in the Afternoon saw an Island bearing W.S.W. distant 11 Leagues; then hoisted our Colours, as a Signal to the other Ships. *Friday* 14. at Noon the same Island bore S.W. half W. distant 10 Leagues; and then we saw another large Island, which bore N.W. distant 12 Leagues. Soon after perceiv'd several Spouts, which came very near our Ship, looking like boiling Water, smoaking in a Circle, drawn up into the Air, and when it comes near a Ship, for Want of the Moisture it sucks, breaks, and may be of bad Consequence to the Decks, Masts, Sails, and Rigging, if not prevented. Besides that, it is reputed very unhealthy, by Reason of its hot sulphurous Smell. When these Spouts come near a Ship, we commonly fire Shot to break them, as the *Dutchess* did, the Day before, at two, which were just under my Stern, and broke one of them; for I could not bring any of my Guns to bear upon them, and therefore clu'd up my Top-Sails, and put the Ship before the Wind; which is the best Way to receive least Damage, when they cannot be broke. It commonly proves bad Weather after these Spouts.

Moratay Island. *Saturday* 15. after lying by all Night, made Sail for the N.W. Island, which I suppos'd to be *Moratay*; but it prov'd a Mistake; for another which I then took for *Gilolo*, prov'd to be *Moratay*. This Island is high Land,

Land, in Length about a Degree, and at least 30 Miles broad, lying at the *N. E.* End of *Gilolo*, diftant four Leagues, the *North* End of it, according to my Reckoning, in 2 Deg. 20 Min. *North* Latitude, and Longitude from *Iguana* 17 Deg. 0 Min. *Weft*. Near this Ifland we faw abundance of Sea-Weeds, floating on the Water, which at a Diftance look'd very like Bunches of Hops.

The Ifland of *Gilolo* above-mention'd, is very large, lies under the Equinoctial, and is of a very irregular Shape, having four long Points of Land running out feveral Ways, one of them about 20 Leagues in Length, another about 50. The capital City is call'd *Gilolo*, which is alfo the Name of a Kingdom. The other Towns of Note in it, are, *Cuma, Maro, Tolo,* &c. The Inhabitants are *Mahometans*. The *Dutch*-men Aboard us, who had been before in thefe Parts, faid, there were fome *Dutch* Soldiers in this Ifland, as on all the *Moluccos*, to take Care of the Spice, and cut down all the Trees, excepting fuch a certain Number, as they knew would fuffice to furnifh their own Trade. *Ternate, Tidore,* and the other Spice Iflands, now under the *Dutch,* after their expelling the *Portuguefes* and *Englifh,* almoft join to *Gilolo.*

Gilolo Ifland.

We now endeavour'd for fome Days to get as much as poffible to the *Weftward,* to weather *Moratay,* and found the Current help'd us, by Reafon of its being to the *N. W.* the Winds being variable from the *S. S. W.* to the *W.* by *S.* and no lefs various Weather, fometimes very fqually, wet, and thick, and then again extream hot and calm. I look upon it as the worft of Weather, and very unwholfome, and may perhaps be occafion'd by the
Clouds,

Clouds, which are drove some thousands of Miles, by the Trade-Winds, and stop here, as if it were their Place of Rendezvous, then meet with opposite Breezes, which are here call'd *Monsons*, and drive them back; which contrary Motion and Agitation sets on Fire the sulphureous and nitrous Matter, and breaks through the Clouds in Thunder and Lightning, to such a Degree, as this Part of the World exceeds all others I have been in, and sometimes it is very dreadful.

Monsons. We continu'd 'till the 23d of *April*, plying to Windward, hoping to weather the *Westermost* Point of *Moratay*, with such terrible *Westerly* Winds, and Storms of Thunder and Lightning, that I could compare it to nothing but Doomsday. Besides what has been said above, I attribute this to the shifting of the Monsons; for about the *Molucco* Islands, the *S. E.* Wind begins to reign the latter End of *April*, and continues 'till the latter End of *September*, being reckon'd the bad Monson, because very subject to hard Gales of Wind, with much Thunder, Lightning, Rain, and thick Weather, the most violent Part of it in *June*, *July*, and *August*; after which it abates, and in *September* quite breaks up. Then begins the *Westerly* Monson, which continues the other half of the Year, and is counted healthy, because the Weather is generally clear, with moderate *Westerly* Breezes of Wind. At *Batavia* it is quite contrary; for there the *Easterly* Monson is counted the good, and the *Westerly* the bad; the *Easterly* beginning in *April*, and ending in *October*; and the *Westerly* lasting the other six Months; there being generally thick Weather, with hard Gales of Wind,

Wind, and much Thunder, Lightning, and Rain in *December*, *January*, and *February*.

To return to our Voyage, the Weather prov'd such, as broke most of our Main Shrouds, several of our Stays, most of our running Ropes, and the Mizen Gears; so that the Yard came by the Board, broke in several Pieces, and knock'd down Mr. *Pope*, my Lieutenant, who lay speechless for some Time, but soon after came to himself. Most of our Sails were split, and I bent others. I was supply'd with new Shrouds by Capt. *Rogers*, and with others from Capt. *Courtney*; besides all which, my Ship was very leaky. The *Duke* and *Dutchess* far'd not much better, and the Prize split most of her Sails, but the Ship is so strong, that no Weather could well damage her. I could not imagine we should have met with such boisterous Weather in that Latitude, so near the Sun. It was as bad as what we met with at Cape *Horn*, but only warmer. Sometimes the excessive hard blowing oblig'd us to lie by, and then again to bear away, by which we lost Ground considerably. *Foul Weather.*

Monday 24: we had the Wind at *W. S. W.* and *S. W.* then stood to the *Northward* for the Island *Mindanao*, the bad Weather continuing 'till *Friday* 28, when we had less Wind at *South*, and fine Weather, when I got some salt Provisions from the *Duke* and *Dutchess*, mine being almost spent, and the Allowance so short, that I have known the Men give a Groat or six Pence a-Piece for Rats, and eat them very savourly. Latitude this Day 5 Deg. *North*, Longitude from the Island *Moratay* 2 Deg. 19 Min. *West*.

Saturday,

Saturday, April 29. our Water and Provisions being very short, and the *Duke* continuing leaky, a Committee was held on Board the *Batchelor*, where Capt. *Rogers*, and others, were for making directly to some Port in the Island *Mindanao*; but it was carry'd against them for several Reasons. This Island is one of the *Philippines*, and the largest of them, next to that of *Luzon*, where is the *Spanish* capital City of *Manila*. The Compass of *Mindanao* is about 300 Leagues, and the Body of it lies in about seven Degrees of *North* Latitude. The Inhabitants are most Gentiles about the Mountains up the Inland, and *Mahometans* about the Sea-Coast, except on the *North* Side, where the *Spaniards* have subdu'd a considerable Part, and converted many to Christianity; the King of *Mindanao*, who lives up the Country, being tributary to *Spain*. The *Mahometans* know very little of their Religion, and the Idolaters are a brutal Sort of People. There is on the Mountains a Breed of perfect Blacks, who are scarce a Degree above Beasts. These, and many others of the Natives, go stark naked, and delight in being below Men, for the so much admir'd Sake of Liberty. Cinamon grows here wild on the Mountains, which also afford Gold; for it is found not only in the Rivers, but very often by digging; and in the Sea they take very good Pearls. The Woods are stock'd with Variety of Birds, and several Sorts of Beasts running wild.

Having mention'd the *Philippine* Islands, it would be improper to pass by, without giving some short Account of them. They are so many, that their Number cannot easily be ascertain'd, lying from 5 to 20 Degrees of of *North* Latitude; not to extend them, as some have done, to include *Celebes*, and many others, which cannot properly belong to them. The greatest of them, are *Luzon*, *Mindanao*, *Tandaya*, *Mindoro*, *Masbate*, *Panay*, *Isla de Negros*, *Abuyo*, *Cebu*, and *Matan*, not to descend to the many small ones. They lie S.E. from *China*, N.E. from *Borneo*, and *West* from the *Ladrones*, or *Marian* Islands. The largest of them, and chief Seat of the *Spaniards*, to whom most of them are subject, is *Luzon*, where is the City *Manila*, their Capital. *Michael Lopez de Legaspi* subdu'd this Island, in the Year 1543; others were afterwards reduc'd by Degrees, and the last Conquests were in that of *Mindanao*, which *Don Sebastian Hurtado de Corcuera*, Governor of the *Philippines*, began to reduce in the Year 1635. I will not enter upon any Discourse concerning the Natives, or particular Descriptions of the several Islands, which would take up more Time than can be spar'd from our Voyage; but Trade being the Support of all Nations, and we having the *Manila* Ship in our Company, it will be convenient briefly to mention what the Commerce between these Islands and *New Spain* consists in. The natural Commodities of the Islands themselves, are Pearls taken in the Sea round about them, Amber-Grease found also in the Sea; excellent Civet taken from the Cats running wild on the Mountains, which they catch with Snares; Cinamon growing wild on the Mountains of *Mindanao*; Wax made in many of the Islands; a Sort of Cotton Cloth call'd *Campotes*, wove and much worn by the *Indians*; and a considerable Quantity of Gold, found yearly in the Rivers, and brought down from the Mountains. All these

Trade of those Islands.

these Commodities are transported to *New Spain*, and with them many more they have from *China*, as Callicoes and Muslins, some of them the finest in the World; all Sorts of wrought Silks, and great Quantities of Silk both raw and spun, Silks flower'd with Gold and Silver, with Loops and Galoons, Cusheons, Canopies, and Purelane; also Pearls, Gold, Iron, Thread, Musk, curious Umbrellas, sightly false Jewels, Salt-peter, Paper, white and of several Colours, Japan Work, and inimitable emboss'd Works. The main Return from *New Spain*, for all these Goods, is Plate, the only Thing the *Chineses* covet, who trade at the *Philippines*, and some *European* Commodities, for the Use of the *Spanish* Inhabitants of those Islands; which we will now leave, to return to our Voyage.

CHAP.

CHAP. IV.

The Voyage continu'd among the Islands of India, and through several Streights to Batavia, the Capital of the Dutch Dominions in those Parts; some Particulars of the Island, Bouro, Cambava, Wanshut, Buton, Solayo, Madure, Carimon Java, and the General's Island.

At a Committee held on Board the *Batchelor*, April 29. 1710.

IT is agreed to make the best of our Way to the *Island* of Talao, *where we hope to supply our selves with Wood, Water, and Provisions, to cruize* 10 or 12 *Days for the same. In case the Wind should present sooner, so as that we can fetch* Ternate, *then to make the best of our Way for it; and in case the Wind should not present for* Ternate, *or the Island of* Talao, *then, if we see Occasion, to make the best of our Way for some Port of* Mindanao. *For all Opportunities of Slatches, going about, and carrying the Light,* &c. *we leave it to Capt.* Courtney *and Capt.* Dampier.

Sign'd by most of the Committee.

Tuesday, May 2. 1710. in the Morning saw a large Ring, like a Rainbow, quite round the Sun. We had often a Ring or Bur about the Moon, and seldom miss'd of hard blowing Weather soon after. Latitude 3 Deg. 30 Min. *North*, Longitude from the Island *Moratay* 2 Deg. 55 Min. *West*. We ran by the

A Ring about the Sun.

34 *A* Voyage *to the* South Sea, *and*

Island of *Talao*, without seeing it, by Reason of our being to the *Westward*; and *Wednesday* 3. in the Morning saw some Land, which we took to be a small Island lying between *Celebes* and *Gilolo*, bearing *W*. by *S*. distant 10 Leagues. *Thursday* 4. Latitude at Noon 1 Deg. 50 Min. Longitude from *Moratay* 2 Deg. 55 Min. *West*.

Strong Current. *Monday* 8. having had very uncertain Weather for some Days past, we now perceiv'd a very strong Current had set us to the *Eastward* near five Degrees more than we expected; for, to our great Astonishment, we found the Land we had seen the *Thursday* before, was *Moratay*, because, having made the Land this Morning, it prov'd to be Cape *Noba*, a Promontory at the *East* End of *Gilolo*, bearing *S. S. E.* distant 15 Leagues. At the same Time two other small Islands bore, the one *South*, and the other *West*. Perceiving now that we could not get up to *Ternate*, which we thought to have done with Ease, resolv'd to make the best of our Way thro' the Streight of *Gilolo*, that Passage lying between Cape *Noba* and the *Western* Point of *New Guinea*, or the Land of *Papous*, where there are many Islands to the *Eastward*, and some of them very near *Gilolo*. Latitude this Day at Noon *per* Estimation 0 Deg. 30 Min. *North*. I reckon Cape *Noba* lyes in the Latitude of 0 Deg. 5 Min. *North*, and Longitude from *Moratay* 2 Deg. *East*.

Wednesday, May 10. found a strong Current setting us to the *West*; and having had little or no Wind, gain'd little Ground; however got very near in with the Bite of *Gilolo*, the Land at Noon bearing from the *S. E.* to the *S. W.*

Cape

Cape *Noba.*

When it bears *S. S. W.* distant nine Leagues, shews thus.

The low flat Islands bearing *N. N. E.* from Cape *Noba*, distant 14 Leagues, and in about 20 Min. of *North* Latitude, shew thus when they bear from you *N.* by *E.* distant two Leagues. They are very full of green Trees.

Another Island bearing from Cape *Noba*, *E.* by *S.* distant 18 Leagues, shews thus when bearing *N. E.* by *E.* four Leagues distant from you. When you are farther to the *East*, you'll see several other small ones, almost joining to it.

The high Land bearing *E. S. E.* from Cape *Noba*, distant 18 Leagues, and in 20 Min. Latitude *North*, shews thus when it bears from you *E.* by *N.* distant six Leagues. *South* from it, run out several small Islands or Rocks. Some Part of this Cape shew'd barren about the Hills, but coming near the Vales, appear'd full of green Trees, and very pleasant.

An Island bearing *S. S. W.* distant 30 Leagues, and in the Latitude of 1 Deg. 30 Min. *South*, shews thus when bearing from you *S.* by *W.*
distant

distant five Leagues, being a small high Island, and very woody. To the *N.W.* of it, are two small Islands at some Distance, and one to the *S.E.*

Monday, May 15. The Winds having been very variable from the *South* to the *East*, and *E.S.E.* and a strong Current still setting to the *Westward*, we gain'd the Days past little Ground; but pass'd by several Islands, lying near Cape *Noba*, one of which lyes in 20 Min. *North* Latitude, and bears *N. N. E.* from the said Cape, distant 14 Leagues, mention'd p. 35, and laid down, being the second of the Bearings. It is a low flat Island, very well cloath'd with green Trees, and affords a pleasant Prospect; but at the *East* End of it there are Shoals, and at the *West* End is another small round low Island, almost joining to it. We sent our Pinnaces a-shore at this Island, to see if we might wood and water, and whether there were any Inhabitants. At their Return, they inform'd us there were none to be seen, but only Places where there had been Fire lately, no Water, yet abundance of large Trees of several Sorts, and among them many Cabbage-Trees, and Plenty of Fish near the Shore. We stood to the *East* towards several other Islands, some of which lye near the Latitude of the Cape, and are higher Land, as appears by the Figures above-mention'd, with deep Water about most of them; for we sounded several Times, when clear of the flat Island, and found no Ground. Our Men at that Island saw the Track of several Tortoises, but could get none. We ply'd to Windward, and weather'd the Cape the 14th in the Morning, and stood away *S. S. E.* and the Current setting strong to the *Westward*, found

we

we made but a *S. S. W.* Course. That same Evening we saw the Land of *Papous* or *New Guinea*, which is high craggy Land, inhabited by Blacks; but we were a good Distance from the Shore. This Day the *Duke*, *Dutchess*, and *Marquis*, were supply'd with near a Month's Bread from the *Batchelor*; our Water grew very short, so that the Men had but a Quart a Day for some Time past. This 15th Day at Noon, Cape *Noba* bore *N. N. E.* distant 19 Leagues; at the same Time saw an Island bearing *S. S. W.* 8 Leagues from us. It is small and high, and I take its Latitude to be 1 Deg. 30 Min. *South*, mention'd p. 35. and represented, being the 5th of those Bearings; *N. W.* from it, are two other small ones, and another to the *S. E.* Our Latitude this Noon *per* Estimation, 50 Min. *South*; Longitude from Cape *Noba*, 20 Min. *West*.

Monday 17. the Weather excessive hot, with small Breezes from the *N. E.* to the *E. S. E.* and sometimes calm, the Current still continuing to set very strong to the *Westward*. At Noon got up to the Island above-mention'd, in 1 Deg. 30 Min. *South* Latitude, which is full of green Trees from the Water-side, up to the very Mountain. I sounded very near the Shore, and found no Ground with the deep Sea-Line. From hence we could see several other small Islands, full of Trees, lying, as I thought, in a Line from this Place *N. W.* to *Gilolo*, and *S. E.* to the Coast of *New Guinea*. Naturalists differ very much in Opinion concerning these Islands of *India*, some affirming they were created with the World, when the Author of Nature separated the Land from the Water; others, that they were made by *Noah*'s Flood; and others suppose

pose them to have been cut off from the Continent by Inundations of Provinces, Tempests, Earthquakes, Eruptions of Fire, and other Accidents, which occasion Alterations both at Sea and Land; and this seems probable enough, for where shall we hear of more burning Mountains, dreadful Earthquakes, amazing Thunder and Lightning, terrible Hurracanes, strong Currents, and violent Rains, than in these Parts, at certain Seasons of the Year? These are forcible enough to rend one Piece of Earth from another. This Day at Noon we were in 1 Deg. 25 Min. Latitude *South*, and 34 Min. Meridian Distance from *Noba West*. At Midnight made Shift to get through between the Islands, where we met with strong Tides and Counter-Currents, insomuch that sometimes our Ship would not feel the Helm, but ran quite round.

Thursday 18. made the *Southern* Part of *Gilolo*, and another long Island, which lies to the *Eastward* of it, and at the same Time saw the Islands we had left the Day before.

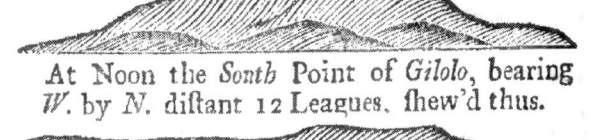

At Noon the *South* Point of *Gilolo*, bearing *W.* by *N.* distant 12 Leagues, shew'd thus.

The Body of the long Island above-mention'd, bearing *N. W.* by *N.* distant 6 Leagues, shew'd thus.

The

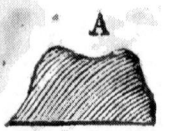

The other three small Islands, shew'd thus; (A) bearing *N. N. E.* six Leagues distant; (B) *N. N. E.* half *E.* seven Leagues; and (C) *N. E.* half *N.* 10 Leagues; the Latitude of this last 1 Deg. 30 Min. as was said before. The Ship's Latitude this Day at Noon 1 Deg. 50 Min. *South*, Longitude from Cape *Noba* 51 Min. *West.*

Saturday, May 20. a fresh Breeze at *S. E.* with strong Currents, setting us sometimes to the *South*, and at others to the *North*. Past by many Islands, leaving them to the *S. E.* of us. At Eight this Morning made the Island of *Ceram*, as we then thought, but it prov'd to be *Bouro*, and were come very near by Noon. It is high Land, full of Hills and Vales, and shews woody, but the Trees did not look so green as in the small Islands. This is an oval Island, inhabited by a People much like those of *New Guinea*, and are *Mahometans*; produces Plenty of Rice, some Spice, and other *Indian* Commodities. This Island *Bouro*, when the *Eastermost* Point bears *E.* half *S.* distant six Leagues, and the *Westermost* Point *W. S. W.* distant nine Leagues, shews as over the Leaf, at p. 40.

Bouro Island.

At Noon the *Eastermost* Point of it bore E. half S. distant six Leagues, and the *Westermost* W. S. W. distant nine Leagues, in which Position it shew'd thus. Ship's Latitude three Deg. *South*, Longitude from Cape *Noba* to the *Western* Point of *Ceram* 1 Deg. 30 Min. *West*.

Sunday, May 21. having a hard Gale of *Easterly* Wind at Night, stood to the *Westward*, and for some Time had such a Head-Sea, occasion'd by the Current, that I thought the Ship would have pitch'd some of her Masts by the Board, which made her so leaky, that we could hardly clear the Water. *Monday* 22. in the Morning little or no Wind; whence I conclude *India* is the worst Country in the World for Storms and Calms, and thick, rainy, unwholsome Weather. Latitude at Noon 3 Deg. *South*, Longitude from the *N. W.* Point of *Bouro* 30 Min. *West*.

Thursday, May 25. at Four in the Morning, steering away *S. W.* saw a low Island right a-head of us; and when it was clear Day, bore away for the *N. W.* End, where we saw an opening, and prov'd to be two Islands, almost join'd. They were very full of green Trees, and by the Sea-side several Groves of Coco Nut, Plantan, and other Sorts of Fruit Trees, which appear'd very pleasant. Up the Bay we saw several Boats, Houses, and abundance of the Native *Malayes*, walking along the Shore. We sent in our Boats for Provisions and Pilots, and the *Duke* and my self turn'd up very near to the Town; but sounding several Times, found no Ground. The Natives inform'd us there was a Bank opposite to the

Town,

Town, where we might anchor. Abundance of People came off with *Indian* Wheat, Coco Nuts, Yams, Patatas, Papas, Hens, and several Sorts of fine Birds, to truck with us for Cloaths, Knives, Sciffars, and other Toys, being very civil to all Appearance. They are *Mahometans*, of a middle Stature, and tawny; but the Women somewhat clearer than the Men, having very long black Hair, their Mouths, Lips, and Noses small. They wear a Linnen Waftcoat, which reaches only to the lower Part of their Breafts, and about their Wafte a Piece of Cloth three or four Yards wide, and a Yard deep, which they wrap about them inftead of a Petticoat. The Men that came off, were all naked, having only a Cloath roll'd about their Middle, to cover what ought to be. Some of the better People had a loose Sort of Waftcoat, and a Piece of Linnen roll'd about their Head, with a Cap of Palm-Tree Leaves to keep the Sun from scorching. They brought off several Cacatoes and Parrots, very fine Birds. Along the Shore-side we saw several Weares they had to catch Fish. In turning up, we found the Current very ftrong againft us, and the Prize loft Ground confiderably; wherefore in the Evening the *Dutchefs* fir'd a Gun. We ran out, and drove all Night. The Names of these two Iflands, are, *Cambava* and *Wanfhut*.

The which

when

42   *A* Voyage *to the* South Sea, *and*

when bearing S. S. W. half a League diftant, fhew as in the Margin. From hence the Middle of the Ifland *Buton* bears *Weft*, diftant 8 Leagues. I take thefe two Iflands to be in the Latitude of 5 Deg. 10 Min. *South*, Longitude from the Ifland *Bouro* 2 Deg. *Weft*.

Sunday 28. at Four in the Afternoon came to an Anchor in 18 Fathom Water, 3 Leagues from the Town of *Buton*, in 5 Deg. 40 Min. Latitude *South*, Longitude from *Cambava* 30 Min. *Weft*, off a Point of *Pulo Shampo* on the *N.* by *E.* Side, *Pulo Shampo* bearing *S.* by *W.* diftant half a League; *Pulo Paffia N. N. W.* two Miles; *Pulo Bouna N.* by *W.* five Leagues; the high Land off of *Pulo Cubina W.* eight Leagues, and the *Southermoft* Land in Sight *W.* by *S.* 10 Leagues. *Monday* 29. in the Morning Capt. *Dampier*, Mr. *Connely*, and Mr. *Vanbrug* went with a Prefent to the King of *Buton*, to defire he would fupply us with Provifions, and a Pilot to conduct us to *Batavia*. *Tuefday* 30. in the Morning a Parao came from the King, with a Nobleman, who had neither Shoes nor Stockings, and a Pilot to carry us up to the Town. He afk'd, how we durft come to an Anchor there, without Leave from the Great King of *Buton*, as he ftil'd him. He brought each Commander a Piece of *Buton* ftrip'd Cloth, a Bottle of Arrack, fome Rice in Bafkets, *&c.* as a Prefent from the King; as alfo a Letter from the Officers we had fent afhore, giving an Account that they had been very well receiv'd, and that the Town where the King refides, is large, wall'd and fortify'd, and has feveral great Guns. Another Prefent was return'd, and five Guns fir'd by every Ship at the Meffenger's going off; at which he feemed ed very well pleas'd. We wooded and watter'd at the Island *Shampo*, and several *Paraos* came off to us with Fowl, *Indian* Corn, Pompions, Papas, Lemons, *Guinea* Corn, &c. which they truck'd for Knives, Scissars, old Cloaths, &c. The People were civil, but sold very dear; yet, our Officers making a longer Stay at the Town than was intended, we began to suspect they might be detain'd, those *Moors* being very treacherous. However, we heard from them every Day, and on *Sunday*, *June* 5. the *Dutchess*'s Pinnace came down with Lieutenant *Connely*, who told us there was four Last of Rice coming, which was bought of the King, and cost 600 Dollars, 50 Dollars in Tale being allow'd to make up the Weight, because the Royals were light, and that Mr. *Vanbrug* was detain'd for the Payment. The next Morning it came, and was equally distributed among the four Ships, some great Men coming to deliver it, and receive the Money. A *Portuguese* sent by the King, was detain'd 'till our Boat returned, and the Provisions began to come more plentifully and cheaper.

The Town of *Buton* is seated on the Ascent of a Hill, on the Top whereof is a Fort, enclos'd with an old Stone Wall, on which there are Guns and Pedrero's mounted. The King and a considerable Number of People live in the said Fort, where an Herb Market is kept every Day. The King has five *Wives*, besides Concubines, and four Men call'd *Pury Bassas*, who carry great Canes, with Silver Heads, to manage his Affairs. His Majesty, on his long black Hair, wears a Sort of green Gause strew'd with Spangles, goes always bare-footed and bare-legg'd, is sometimes clad

Buton Town.

like

like a *Dutch* Skipper, but when he appears in State, has a long Calico Gown over his short Jacket. In Council he sits on a Chair covered with red Cloth, is always attended by a Serjeant, and six Men with Match-Locks, besides three others, one of which wears a Head-piece, and carries a large Scimiter in his Hand, another holds a Shield, and the third a great Fan; four Slaves sit at his Feet, one of them holding his *Betele* Box, another a lighted Match, another his Box to smoke, and the fourth his Spitting-Bason. The petty Kings and great Men sit on his Left Hand, and before him, every one of them attended by a Slave in the Council-Chamber, where they smoke Tobacco, and chew *Betele* in the King's Presence, and speak to him sitting cross-legg'd, joining their Hands, and lifting them up to their Forehead. The Town of *Buton* is very populous, and by it runs a fine River, which they say comes down from 10 Miles up the Country, ebbs and flows considerably, and has a Bar at the Entrance, so that Boats cannot come out at low Water. At least 1500 Boats belong to this River, 50 whereof are *Paraos* for War, carrying Pedreros, and 40 or 50 Men each. About 50 Islands are tributary to the King, who sends some of his *Paraos* once a Year to gather in the Tribute; which consists of Slaves, each Island giving him two Inhabitants out of every hundred. There is one *Mosque* at *Buton*, which is supply'd with Priests from *Moca*, the People being *Mahometans*. They are great Admirers of Musick. Their Houses are built upon Stilts. *Dutch* Money is current there, and *Spanish* Dollars.

Wednesday

Round the World. 45

Wednesday, June 7. 1710. The *Duke's* Pinnace being come from the Town with Mr. *Vanbrug*, and the rest of the Men, at Five this Morning the *Dutchess* fir'd a Gun to unmoor, which I did, and weigh'd ; but there being little Wind, and a strong Tide against us, came to an Anchor again at Noon. The Gunner of the *Dutchess* was sent Aboard me, and another Officer Aboard the *Duke*, for a Mutiny.

Thus shews the Island *Buton*, and other Parts about it, here mention'd, Lat. 5 Deg. 55 Min. *South*. At

46 *A* Voyage *to the* South Sea, *and*

At Four in the Afternoon weigh'd, with a small Breeze off the Shore, and made easy Sail all Night, to keep the Prize Company, steering away *W. S. W.* to get clear of Point *Cubina*; and *Thursday* 8. had a fine Breeze at *East*, and fair Weather. At Noon Point *Cubina* bore *N.* distant 7 Leagues. The Wind then coming to *S. E.* we hal'd up *W.* by *N.* 'till Twelve at Night, and then brought to 'till

Solayo Island.

Friday, June 9. at Four in the Morning, then made Sail again; and as soon as it was Day, saw the Island *Solayo*, or *Zelayer*, as our Seamen call it, lying very close to the great Island *Celebes*, and inhabited by *Malayans*, who are said to pay Tribute to the *Dutch*. Between the *South* End of *Celebes*, and this Island of *Solayo*, are three small low Islands. The best Passage is between that which lies next to *Solayo*, and a little one lying to the *Northward* of that, as may be seen by this Figure. This is call'd the second Passage from *Solayo*, and much the better; for in the first there are many Shoals, which may be seen at a Distance, and in that next to *Celebes* it is necessary to anchor sometimes for the Land-Wind; but there is no Danger in going

going thro' this, provided there be a leading Gale of Wind, and you mind to keep over to the Ifland that lies to the *Northward*, becaufe there commonly fets a ftrong Current from the Shore of *Celebes* to the *Southward*. It is very dangerous going to the *Southward* of the Ifland *Solayo*, and the Pilot affur'd me, that the *Dutch* will never venture that Way; but that all the *Amboyna* and *Ternate* Ships come through the aforefaid Paffage. This Morning we faw a *Parao* off the Paffage, and gave Chafe to her 'till Twelve, then brought to for the Prize, and the Pinnaces went in Queft of the *Parao*, and foon brought the Mafter Aboard the *Dutchefs*. He was a *Malayan* belonging to *Macaffar*, bound thither, came laft from *Buton*, and inform'd us, that he had been Pilot to feveral *Dutch* Veffels in thofe Parts, and engag'd to carry us thro' thefe and the Streights of *Salango*, which was very agreeable, we being all unacquainted there. We promis'd him a Suit of Cloaths, and as much Money as he could in Reafon afk, to go with us to *Batavia*. By Night we were through the Paffage, without any Difficulty. Our Courfe corrected from the Ifland *Shampo*, to the Paffage of *Solayo Weft*, 100 Miles. Latitude *per* Obfervation 5 Deg. 45 Min. *South*, Longitude from *Shampo* 1 Deg. 40 Min. *Weft*.

Thus

Thus shews the Passage of *Salango*, when the Islands bear from you, as on the Side.

Saturday 10. in the Afternoon we steer'd *N. W.* by *N.* between the *S. W.* Part of the Island *Celebes*, and *Salango*, an Island lying off, sometimes haling up *North*, and saw the Water was chang'd, and we were in Soundings most Part of the Day. At Two in the Afternoon sounded, and had always six or seven Fathom Water at least, in running thro'. At Six got clear, and stood to the *Westward*, to go between two small Islands, which lay to the *Westward* of us; but Night drawing on, oblig'd us to come to an Anchor, in 12 Fathom Water, the *S. W.* Part of the Island *Celebes* bearing *N. E.* by *N.* distant five Leagues, and the Passage between the two Islands *S. W.* by *S.* distant two Leagues.

Sunday 11. at Six in the Morning weigh'd, and steer'd away *S W.* by *S.* through the Passage, and then more to the *West*, having a fresh Breeze of Wind at *S. E.* Fifteen Leagues *S. W.* by *W.* from this Passage, are three small low Islands, and near to them there are Shoals; to avoid the which, all Ships keep to the *Southward*,

ward of them, and then hale away more *We-sterly*. At Noon the *S. W.* End of the Island *Celebes* bore *East*, distant 10 Leagues. Near that Place the *Dutch* have a strong Garrison, call'd *Macassar*, which formerly belong'd to the *English*. Our Pilot's Boat attended us 'till we were through the last Streight, and then bore away towards *Macassar*, without going Aboard the *Dutchess* for their Master; with which we were well pleas'd; as being secure of so able a Pilot; and he no less contented, expecting a good Reward. Our Latitude *per* Observation at Noon 5 Deg. 45 Min. *South*, Longitude from the Streights of *Solayo* 1 Deg. 30 Min. *W.* the *S. W.* End of *Celebes* bearing *E.* distant 10 Leagues. In the Afternoon saw 3 small Islands to the *Northward*; at Four one of them bore *N. N. E.* distant seven Leagues, and another *N. N. W.* distant four Leagues. 'Till that Time we had steer'd *S. W.* by *W.* and then all Night *W. S. W.*

Monday 12. The last twenty four Hours our Course was *W. S. W.* Distance 110 Miles. Latitude at Noon *per* Estimation 6 Deg. 27 Min. *South*, Longitude from the *S. W.* Point of *Celebes* 2 Deg. 12 Min. *West*. At Six in the Evening a small Island bore *N. W.* by *W.* distant 11 Leagues; and *Tuesday* 13. at Six in the Morning, another small Island bore *S. S. E.* distant nine Leagues. The last twenty four Hours sail'd *West*, half *South*, Distance 110 Miles. Latitude this Noon 6 Deg. 38 Min. *South*, Longitude from the *S. W.* Point of *Celebes*, 4 Deg. 1 Min. *West*. In the Afternoon made an Island lying to the *Eastward* of *Madure*, and at Six in the Evening the *West-stermost* Part of it bore *S. W.* by *W.* distant seven Leagues.

Wednesday 14. saw several *Paraos*, plying to Windward, and Fisher-mens Buoys, as we ran along the Shore. A very sweet Scent came off from the Island, which shews low and pleasant, we being then off *Madure*, a large Island, lying at the *N. E.* End of *Java*. At Noon the *N.* End of *Madure* bore *S.* distant 12 Leagues. Our Course the last 24 Hours *W.* Distance 100 Miles. Latitude *per* Estimation 6 Deg. 38 Min. *S.* Longitude from the *S. W.* Point of *Celebes*, 5 Deg. 41 Min. *W.* Steer'd *W.* 'till Ten at Night, then *W.* by *S.* 'till

Thursday 15. at Six in the Morning, then hal'd away *W.* by *N.* and *W. N. W.* This Morning saw a high Land, being the *Northermost* Part of the Island *Java*, call'd *Japara*. Thus it shews when the Island bears from you *W. S. W.* as in the Margin, distant five Leagues. At Noon the *Westermost* Part of *Japara* bore *W.* distant 7 Leagues. Latitude *per* Estimation 6 Deg. 38 Min. *S.* Longitude from the *S. W.* Point of *Celebes*, 7 Deg. 11 Min. *W.* Course the last 24 Hours *W.* Distance 90 Miles. From Noon steered *W. N. W.* and *N. W.* by *W.* 'till Six in the Morning.

Friday, June 16. then saw the Island *Carimon Java*, which shews as in the Margin, when bearing *N. N. E.* distant 4 Leagues. Then hal'd away *W.* and *W.* by *S.* The last 24 Hours Course *W. N. W.* Distance 90 Miles; Latitude *per* Estimation, 6 Deg. 4 Min. *South*; Longitude from the *S. W.* Point of *Celebes*, 8 Deg. 34 Min. *West*.

Monday

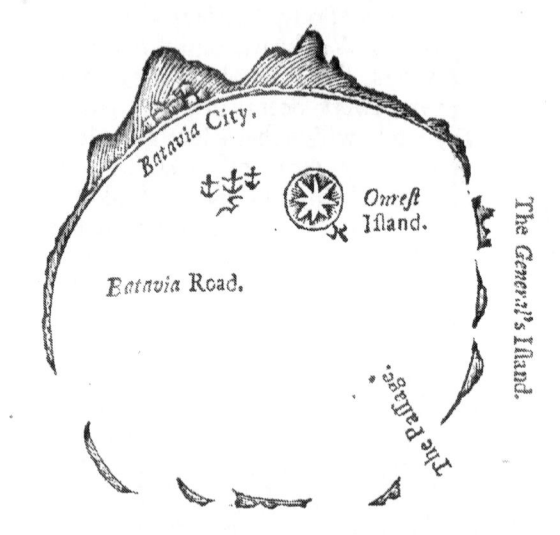

Monday 19. The Days not mention'd, had Java Land and Sea Breezes, from the *N. N. E.* to Island. the *S. E.* and very hot Weather. Saw two Ships, and several Boats, and pass'd by several Factories. The Island *Java* is in some Places very low Land. We anchor'd several Times two Leagues off the Shore in 15 Fathom Water, very holding Ground. At Seven this Night came to an Anchor in 15 Fathom, about three Leagues *E. N. E.* from the *General's* Island, which lies off the Harbour of General's *Batavia*. On this Island is a Wind-mill, and Island. great Store-houses, where the *Dutch* land the Spice they bring from the *Moluccos*.

Tuesday 20. in the Morning sail'd again, with the Wind off the Sea, and at Noon the *General's* Island bore *W.* by *S.* distant two Miles; whence we could plainly see the Ships in the Road of *Batavia*. *Note*, That when you come from the *Eastward*, you will see

E 2 three

52 *A* VOYAGE *to the* South Sea, *and* three small Islands, one whereof is the *General's*, by which you will know the Harbour of *Batavia*. Our Course since the 16th to the *General's* Island, *West*, Distance 120 Miles. Latitude of the *General's* Island, *per* Estimation, 6 Deg. 4 Min. *South*; of the City of *Batavia* 6 Deg. 10 Min. *South*. Longitude from the *S. W.* Point of the Island *Celebes*, to *Batavia*, 11 Deg. 34 Min. *West*.

CHAP. V.

How a Day is gain'd or lost in sailing round the Globe; a short Account of the Road and City of Batavia; *Victualling and refitting there; Distribution of Plunder; Money advanc'd to Officers; the Ship* Marquis *sold; Orders and Resolutions of the Committee.*

A Day lost in sailing round.

WE came to an Anchor at Seven in the Evening, in *Batavia* Road, on *Tuesday*, *June* 20. 1710, according to our Reckoning, but with the *Dutch* it was *Wednesday*, *June* 21; for we had lost 18 Hours in going round to the *Westward*, and they had gain'd six in sailing to the *Eastward*, which made a whole Day Difference between our Account and theirs. The Reason of it is, that a Ship sailing to the *Westward*, and so following the Course of the Sun, makes every Day something longer than it would be, continuing upon the same Meridian. Thus in every 15 Degrees she removes *Westward*, from the Meridian

dian where she first set out, she gains an Hour, in 90 Degrees six Hours, and in the 360, which compose the whole Circumference of the Globe, will find a whole Day short in her Reckoning, according to the Account of the Place she arrives at. The contrary happens to the Ship that goes to the *Eastward*; for as she advances against the Course of the Sun, she loses so much of every Day, which is thereby shorten'd, and becomes less than 24 Hours, by Consequence gaining an Hour in every 15 Degrees, and in sailing round the World 24 Hours, and therefore will be a Day before the Account of the Place she arrives at. By this it appears, that the Ship which sails round the World *Westward*, loses a whole Day, and that which performs the same Voyage to the *Eastward*, gains a Day. So we having made the greater Part of the Circumference, and the *Dutch* at *Batavia* the other Part the contrary Way, our Loss and their Gain made up the 24 Hours, and thus we came to differ a Day.

We anchor'd here in five Fathom Water, the Ground so soft and ousy, that the Anchor sinks above a Fathom, so that it cannot foul, and therefore Ships always ride single. The Town bore S. by E. distant a Mile and a half; the *General's* Island N. N. E. distant about three Leagues; and the Island *Onrest* bore N. W. by N. distant two Leagues and a half. At this Island the *Dutch* clean and careen all their Ships, and have two Windmills on it to saw Timber. They hale their Ships along the Side of a Wharf, where there are two Cranes to discharge them, and Store-Houses to lay up the Goods. The *Dutchess* fir'd 13 Guns to salute the *Dutch* Flag; but it being Night, he did not then answer, yet the next Morning he sent his Boat Aboard, to beg the Captain's Pardon for that Omission, which he would then repair. Soon after the *Duke* fir'd 13 Guns, and the *Dutch* Flag answer'd both our Ships Gun for Gun. Between Twelve and One, two *English* Gentlemen came Aboard us, the one Captain of an *English* Ship, there being three and a Sloop in the Road, all belonging to *Madrass*. All we Commanders went ashore, and landed at *Bomb*-Key, whence we proceeded to the *Shabander*, who conducted us to the Castle, before *Abraham Van Ribeck*, General of *India*, who receiv'd us very kindly, and ask'd several Questions relating to our Voyage, which we answer'd, shew'd him our Commissions, and ask'd Leave to victual and fit our Ships. He directed us to send him an Account in writing of all the Particulars we stood in need of, and he would give us his Answer.

Batavia City.

We went thence to see the City of *Batavia*, which is the Metropolis of the *Dutch* Dominions in *India*, and seems to be bigger than *Bristol*. It stands in a Bottom, and therefore is not very healthy, but always extraordinary hot. The Inhabitants are *Dutch*, who are Masters of it, a Number of *Portuguese*, and above 200000 *Chinese*, besides *Malayans*, *Javans*, *Persians*, Blacks, &c. The Languages generally spoken among the *Europeans*, are either *Dutch*, *Portuguese*, or *Malaye*. The Houses are large and pleasant, as are the Streets, through the Midst of most of them runs a large Canal, with Trees on both Sides, to shade their Houses, and Arbours. The Town is wall'd and moated round, and well fortify'd with good Guns. The Governor has 10 or 12 prime Men for his Assistants, most of whom

whom have been Governors of *Amboyna*, *Banda*, or some other Places. There are seven Churches in the City, belonging to the *Dutch*, *Portuguese*, and *Malayans*. Thus much may suffice concerning it in general, there being many particular Descriptions of it extant, to which the Curious may have Recourse. I shall only add, that the principal Inhabitants have, without the City, very pleasant Country Houses, Gardens, and Canals. They have of late planted abundance of Coffee-Trees, and have very many Sorts of Fruit. The chief Product of the Island *Java*, on which this City stands, is Pepper.

Soon after our Arrival at *Batavia*, went about fitting the *Marquis*, that being first order'd upon the Careen, the *Shabauder* having allow'd us several *Malaye* Caulkers. When we came down to the Bends, found them, as well as the Stern and Stern-Post, so much worm-eaten and rotten, the Ship being very old, and having only a single Bottom, that we order'd a Survey of Carpenters to view her, who all agreed there was no fitting of her in that Place for going about the Cape of *Good Hope*, her Condition being extraordinary bad, which oblig'd us to hire a Vessel to take out her Lading: Then apply'd our selves to fitting of the other Ships, could not prevail for Leave of the Government to repair to the Island *Onrest*, but were allow'd to go to the small low Island *Horn*, which is near the other, inhabited by a few *Malaye* Fisher-men, and on it are abundance of Coco Nut, Plantan, Papa, Guava, and other Fruit Trees. The Government allow'd us a small Vessel of that Sort they call *Champans*, to careen our Ships by. We then hove down the *Duke* and *Dutchess*,

Onrest and Horn Islands.

efs, and found their Sheathing much worm-eaten in some Places. The *Dutchess*, in heaving down, sprung her Fore-mast, but we soon got another; and the *Duke*, after careening, was still leaky. I repair'd to this Place with the *Marquis*, took in all the Lading of the other Ships, and lay Aboard, on the off Side, to relieve the other Ships when on the Careen. When the Ships were fitted, return'd again to *Batavia* Road, where we rigg'd the three, and sold the *Marquis*, after taking out all her Goods, and most of the Stores, to Capt. *Opie* and Capt. *Oldham*. Then all the Officers and Men were distributed among the other Ships, except one *Dutch*-man, who ran away.

The Weather was extream hot during our Stay. Many Men and Officers fell sick, and I was one of the Number. The Master of the *Duke*, the Gunner of the *Dutchess*, and several of our Men, dy'd of the Flux. *John Read*, a young Man belonging to the *Dutchess*, venturing to swim, had both his Legs snapp'd off by a Shark, which at the second Bite, before we could get him aboard, took off the Bottom of his Belly, so that he was dead before we could take him up. During our Stay, we had the Liberty of the Town and Markets to buy what we pleas'd, yet found it very difficult to get salt Provisions, and were therefore oblig'd to kill several Bullocks, and pickle the Flesh, taking out all the Bones. Arrack, Rice, and Fowl, were cheap, and Beef not above two Stivers a Pound. Several *English* Ships arriv'd here at this Time, as the *Rochester* and *Springer* for *China*, Capt. *Opie* in a separate Stock-Ship, and others of those Parts. The Bay is seldom without some large *Dutch* Vessels,

sels, driving a great Trade from thence to all the Countries round about.

Having continu'd here many Days, and being to settle all Affairs for our Return into *Europe*, several Committees were held to that Purpose, the Resolutions whereof it will be convenient to insert, as containing some material Points relating to our Voyage.

At a Committe held on Board the Batchelor *Frigat,* June 3. 1710, *in the Road of* Batavia.

AGreed to open and new pack all Goods that appear damag'd; other Bails, which were not so, to be new cover'd with waste Cloth, or Tarpaulins, if necessary. That Mr. *Vanbrug* and Mr. *Goodal* should be at every Place, and the rest of the Agents accountable to them, and deliver them Duplicates of the whole, and ready to give Account to a general Committee. That Capt. *Courtney* should provide the Ships with all Necessaries, and Capt. *Rogers*, as soon as well, assist in it, and each Commander to give them a List of what was wanting from Time to Time. Mr. *Pope* to continue ashore, and send Provisions for all the Ships, and keep a Book of the whole, and send it as early as possible, in a Country-Boat, not above nor under 350 Pounds every other Day, or as often as he could conveniently, as also Greens, Carrots, Eggs, or other small Refreshments, more than the common Allowance. A suitable Quantity of Arrack and Sugar to be sent aboard each Ship, to give a Quart *per* Mess; but whilst on the Careen, the Allowance might be enlarg'd, as the Commanders should think fit. If any Thing else were found necessary, for Dispatch, to avoid

Care of Goods and Provisions, &c.

the

the Delays of the Committee meeting, it was left to Capt. *Dover*, Capt. *Rogers*, Capt. *Courtney*, and Capt. *Cooke*, who were to meet and have the same Power as the whole Committee. In Case of such Haste, that all four could not meet, then any three to act; Mr. *Vanbrug* to continue Agent of the *Duke*, Mr. *Goodall* of the *Dutchess*, Mr. *Vigor* constituted Agent of the *Batchelor*, and Mr. *Parker* of the *Marquis*, to keep Account of all Things aboard each Ship, and take Care of the general Interest. Agreed also to divide the Plunder aboard the Batchelor, and Capt. *Dampier* and Mr. *Glendal* appointed Judges of what ought to be given as such, and to govern themselves, as near as possible, to the Resolutions of the Committee of the 13th of *April* 1709. For the more Dispatch, Mr. *Ballet*, Mr. *Appleby*, Mr. *Selkerk*, and Mr. *Smith*, in appraising and dividing the said Plunder, to account for the Officers, and the Sailors allow'd to agree on a Man for each Ship, to act in Conjunction with those above, for the general Good. Mr. *Vanbrug* and Mr. *Goodall* to be present at the appointing and dividing of the Plunder, and to receive what belong'd to their Charge. All Trading was prohibited with the Inhabitants of *Batavia*, or the Natives of the Island, or in any Port of *India*; the same declar'd and publish'd at the Mast of every Ship, and a Protestation made against all Damages that might ensue by the contrary, and the Persons found guilty of such Offence. Resolv'd farther to give 100 Rix Dollars to the Pilot, who brought the Ships from the Streights of *Solayo* to *Batavia*. Order'd, that for promoting of their Dispatch, ten thousand Pieces of Eight should be deliver'd the next Day, being the first

Division of Plunder.

first of *July*, to the Captains, *Dover*, *Rogers*, *Courtney*, and *Cooke*.

July 1. 1710. it was resolv'd in a Committee, that Mr. ——— being found guilty of some Falsities and Mutinies, should be kept in Irons on the *Batchelor's* Poop, under an Awning, during our Stay at *Batavia*, or 'till another Committee should consider farther of it, to prevent his debauching any of the Men.

Officer confin'd.

July 2. 1710. resolv'd to supply the Officers of the *Duke*, *Dutchess*, *Marquis*, and *Batchelor*, with such Sums of Mony as shall be judg'd requisite to furnish them with necessaries for the long Voyage to *Europe*, it being reasonable to suppose that after so tedious a Navigation, without putting into any friendly Harbour, they must be in want of many Conveniencies. Orders were therefore given to the two Agents, Mr. *Vanbrug* and Mr. *Goodal* to deliver the said Sums out of the Money on Board the *Duke* or *Dutchess*, as either Commander should think convenient, &c.

Officers supply'd.

July 20. 1710. agreed in a Committee, that a great Part of the *Marquis*'s Cargo being perish'd through the Weakness of the said Ship, and Damage done by the Worm, which had eaten through her Bends, and a considerable Part of her Bottom, the said Ship *Marquis* should be there sold, and her Cargo distributed among the other Ships. The Captains *Dover*, *Rogers*, *Courtney*, and *Cooke*, empower'd to sell the same.

Marquis to be sold.

September

September 15. 1710. agreed to divide the Money receiv'd of Mr. *Charles Douglas*, for a Quantity of Plate sold him, among the several Ships Companies, which had been adjug'd Plunder. Also to make it our Request to the *Dutch* General, to have leave to careen the *Duke* at the Island *Onrest*, to sell the *Marquis*, to purchase a Supply of ten Hogsheads of *Dutch* Beef or Pork, and for Permission to carry aboard some Arrack and Sugar for the three Ships. It was also agreed to allow the following Particulars for the Use of the Officers in the great Cabbin of every Ship, *viz.* Butter, sweet Oil, Bread or Rusk, Flower, Tamarinds, *Spelman*'s Neap, Cheese, Cape-Wine, and some *Spanish* Money, to buy small Necessaries.

Money to the Seamen.

Request to the General.

Allowance to Officers.

September 23. 1710. *Contents of a small Box adjug'd and accounted Plunder, exchang'd by Order of the Committee, for the Use of the Owners, and valu'd as follows.*

Dollars.

A Box of Plunder. 2 Papers of Gold, making together 57 Ounces, 7 Penny Weight and a half, at 14 *Spanish* Dollars per Ounce ——————————— 0700

1 Paper of Silver, weighing ————— 0032
2 Strings of Stone Rings, set in Gold, and valu'd at 40 *l.* 11 *s.* makes—— 0162,2
A Girdle with Relicks and Toys, at— 0051

42

                                                            Dollars.
42 Boxes, 19 Images set in Silver,⎫
  1 silver Crucifix, 3 Purses, 3 Cups⎪
  tipp'd with Silver, some Amber and ⎬ 0060
  Coral, 2 old Silver Watches, and 1⎪
  Amber Rosary in Filigrane Work-⎭
2 Diamond Rings ——————— 0145
2 Gold Chains ——————————0075
7 Silver hilted Swords ——————0105

      Total 1341,2

 To Mr. *Carlton Vanbrug*,

*W*E *direct you to keep the above Box and Contents, separate from the Plate belonging to the Owners, and to pay the above Sum of one thousand three hundred and forty one Dollars Spanish weighty Money, to Mr.* Robert Fry *and Mr.* William Stretton, *to be divided as Plunder among the several Ships Companies.*

Robert Fry,	Woodes Rogers,
	Stephen Courtney,
William Stretton,	Edward Cooke,
John Ballet.	Charles Pope,

 September 30. 1710: Resolv'd at a Committee, that the following Persons shall proceed from hence for *Great Britain* in the following Ships, Capt. *Edward Cooke*, second Captain in the *Dutchess*, Capt. *Charles Pope*, first Lieutenant in the *Duke*, Capt. *William Dampier* on Board the *Batchelor*; Mr. *John Ballet* on Board the *Dutchess*; Mr. *Robert Knolman*, Master on Board the *Batchelor*; Mr. *Alexander Salcrag*, Master on Board the *Duke*. Agreed also to sell the Ship *Marquis*, with all the Iron, plain and wrought, together or separate, which
           was

Distribution of Officers.

was left to the Management of the Captains *Rogers* and *Courtney*. Also that Captain *Dampier* should be farther supply'd with 200 Dollars *Dutch*.

Resolv'd also by the Committee to supply Mr. *Reynolds* will Money to purchase Provisions and other Necessaries for the Ships, and accordingly 500 *Dutch* Dollars were immediatly put into his Hands, and Directions given for furnishing him with what more should appear to be requisite.

These Resolutions were all sign'd by the Committee, and I have here inserted only the Contents of them, to avoid Prolixity, and proceed on our Voyage.

CHAP. VI.

The Passage from Batavia *to the Cape of Good Hope; the Journal-Table; Letter from thence to the Owners; Description of the Cape, the Town, and Natives of the Country about it,; Preparations to return into* Europe.

Departure from Batavia.

Having compass'd all our Affairs at *Batavia*, and taken in Provisions, &c. to serve us to the Cape of *Good Hope*, on the 14th of *October*, 1710, at Six in the Morning, we were all three Ships under Sail, with the Land Breeze. At One in the Afternoon the

the Sea-Breeze took us, and not being able to weather the Iſlands, were oblig'd to come to an Anchor with our ſmall Bower, in twelve Fathom Water, ouſy Ground, the Iſland *Horn* bearing S. by W. diſtant two Miles, and the *General's* Iſland N. E. by E. half E. diſtant two Leagues. Several of the Supercargoes and Officers of the *Engliſh* Ships in *Batavia* Road, as Capt. *Pike*, Capt. *Opie*, &c. went out with us to Sea, to make up their Accompts, and wiſh us a proſperous Voyage.

The 15th at Six in the Morning weigh'd with a ſmall Breeze at S. W. at Noon had the Wind more *Eaſterly*, *Buffins* Iſland bore N. W. and at Six in the Evening the Point of *Bantam* bore W. half N. diſtant eight Leagues; at Seven came to an Anchor in 14 Fathom, ouſy Ground.

October 16. at Five in the Morning weigh'd with the Land-Breeze, at Noon the Point of *Bantam* bore S. by W. the outermoſt Land W. by S. and the Iſland of *Pulababe* N. N. E. ſeveral ſmall Iſlands on both Sides. Capt. *Opie* and Mr. *Warren* went away to *Batavia*, carrying ſome Plate with them.

October 17. The Wind for 24 Hours paſt was at N. W. and about to S. W. At Noon the Iſland *Duarte* bore N. W. by N. another N. by W. and the outermoſt Point S. by W. We kept plying, making ſhort Trips; Afternoon ſtood in, at Night ſtood off again 'till Nine, then tack'd and anchor'd in 14 Fathom Water, near the *Java* Shore; for when well over towards that of *Sumatra*, we had no Ground. The 18th at Eight in the Morning ſail'd with ſmall Gales of Wind at S. S. E. and S. all Night.

October

October 19. for 24 Hours had a fresh Gale of Wind in the *S.E.* Quarter. At Two in the Afternoon came to an Anchor with our best Bower in a Bay, about a League to the East-ward of *Java* Head, in 15 Fathom Water, ousy Ground, about a Mile from the Shore. Sent our Pinnace for Water, and then our Sail-maker, Wooders, &c. Capt. *Pike* and Mr. *Block* came in a Boat from *Batavia*, the first of them chiefly after his Steward, who I suppose had conceal'd himself, unknown to most of us, aboard the *Batchelor*. In the Evening Capt. *Pike* lent us his Boat and Men; we put into her several of our Men, with Arms and Provisions from each Ship, and sent her away to *Pepper*-Bay, to buy Fowls and other fresh Provisions, giving them for that Purpose Knives, and other Toys, which the Natives there value above Money. In the Evening, much Thunder, Lightning, and Rain, which put us in Fear for the Men sent to *Pepper*-Bay. Continu'd wooding and watering 'till

October 23. 1710. and sent several Men ashore to kill Buffalo's; which being extremely wild, they could shoot none, and durst not stay ashore at Night, by Reason of the many Tygers. One of them was very near seizing a Man of ours, who, to save himself, was oblig'd to take the Water, at least 20 Shots were made at the Tyger before he went off, and they saw several others at the same Time. The *Indian* King, and his People, dealt friendly with us, trucking Fowls, and what else they had to spare, for Knives, and the like. They generally came aboard every Day; and we giving them some Trifle at their going off, they were kind to our Men ashore. The Wind

Wind commonly at *S. E.* a fresh Gale, and we were under some Apprehensions for our Men sent to buy Fowls at *Pepper*-Bay, having heard nothing of them since their Departure, and mistrusting the Boat might have been overset, or the Men detain'd by the *Javans.*

October 24. 1710. *at a general Committee on Board the* Duke.

AGreed to assist Capt. *Richard Pike* with the *Batchelor's* Long-Boat, ready fitted with all Necessaries to carry him to the Port of *Batavia,* he being destitute of a Boat, and desiring our Assistance. Sign'd by most of the Committee.

At a Committee held on Board the Duke, *at an Anchor at* Java-Head, *October 24. 1710.*

AGreed to make the best of our Way to the Cape of *Good Hope*; and if, through any Misfortune, any Ship should lose or part Company in our Passage, either by bad Weather, or otherwise, she to proceed to the Cape of *Good Hope*; and not finding the other Ships there, to stay 20 Days; and the missing Ship or Ships not appearing then, to make the utmost Dispatch to the Island of St. *Helena*; and if not there, to proceed thence, according to the Owner's Orders, for *Great Britain*. Sign'd as usual.

October 25. at Night the Boat return'd, to our great Satisfaction, with the Men, bringing about 12 Dozen of Fowl, some *Mango's, &c.* Capt. *Pike's* Steward came Aboard the *Dutchess,* hoping we would canceal him; but was immediately sent Aboard the *Batchelor* to his Commander, who gladly receiv'd, and promis'd to pardon him.

October 26. 1710. at Seven in the Morning weigh'd with a fresh Gale at *S. E.* at Ten Capt. *Pike* and his Company parted from us, Capt. *Rogers* saluting them with seven Guns; we being far a-head, did not, but sent Capt. *Opey*'s Carpenter, who was conceal'd Aboard of us, unknown to any of the Officers. Then we made Sail, and at Noon *Java*-Head bore *E.* by *N.* distant seven Leagues. From hence to the Cape of *Good Hope*, being a Run at Sea, I shall not enlarge upon it any farther, than as may be seen in the Table adjoin'd, containing all that was observable in this Passage.

It will be needless to add any Thing to this Table, wherein are all material Particulars, as was said above.

Cape of Good Hope.

December 30. 1710. at Nine in the Morning we came to an Anchor at the Cape of *Good Hope*, in seven Fathom Water, red Sandy Ground, the *Norward* Point bearing *N.* by *E.* the Island in the Offing *N.* by *W.* the *Westermost* Point in the Bay *N.W.* half *W.* the Peek *S.W.* by *W.* the Table-Land *S.W.* by *S.* We saluted the Garrison with nine Guns, and they return'd seven. Then the *Donegal India-Man* saluted us with seven, which we also return'd. Some Time after it clear'd up, when the *Duke* and *Batchelor* came in, and saluted with the like Number of Guns. Then we all moor'd *S. E.* and *N. W.* and went ashore to the Governor, who receiv'd us very kindly, and invited all us Commanders and Officers to dine with him at the Fort the next *Sunday*. This Place is very well known for the high Sugar-

Loaf

Loaf Hill, Table-Mountain, &c. of which our Waggoners and other Books give a very good Description; for which Reason I shall not insert any Draughts, as being already common enough, and will only mention what I observ'd my self as to the Town, with some very short Remarks, as to the Country, for the Satisfaction of those who have not met with a more satisfactory Account.

Of the Cape of Good Hope.

THIS Cape was first discover'd by Bar- *By whom* tholomew Diaz, sent out by King John *discover'd*. the 2d of Portugal, with three Ships, about the Year 1486, who being toss'd there by dreadful Tempests, gave it the Name of Cabo Tormentoso, or the Stormy Cape; but at his Return Home, the King hoping then to discover India, chang'd the Name to that of Cabo *Why so* de Boa Esperanza, that is, Cape of Good Hope, *call'd.* which it has ever since retain'd. Some assign Vasco de Gama for the Discoverer of it; but that is a Mistake. It lies in the Latitude of 34 Deg. 15 Min. South, and 17 Deg. East from the Meridian of London. About three Leagues from the Cape, is that the English call Penguin *Penguin* Island, small, but has a Fort on it, and near *Island.* it very good Anchoring; for which Reason Ships often stay there for the Sea-Breeze. Thither the Dutch often banish Malefactors. When they on the Island spy any Ships, they fire Guns, to give Notice to those at the Cape. The said Cape is easy to be known by it's *Marks to* high Table-Mountain, Sugar-Loaf Hill, and *know it.* Lions Mountain, on the two last whereof there are always Look-outs, who fire Guns, and make other Signals, whensoever they see any Ships at Sea. There is another very high Hill, which the *English* call *Crown* or *Charles*'s Mount. This Place is subject to violent Gales of Wind, and in their Summer, which is our Winter, the greatest Storms are generally at S. E. and blow off from *Charles*'s, or the Table-Mountain. Before those violent Blasts come, the Tops of those Mountains begin to be hooded, which the Sea-men call the Table-Cloaths going to be laid, which is thick Clouds appearing on the Tops of those Mountains; and when the Table-Mountain is so hooded all over the Top, it soon blows powerfully.

Storms.

Under that Table-Mountain, in a Bottom, stands the Town, about as big as *Falmouth*, which, as well as the Ships in the Road, is commanded by a strong Fort, standing at one End of the Town, now in the Possession of the *Dutch*, but formerly belonging to the *English*. Wood is somewhat scarce there; but Corn, Wine, Fruit, Sallads, and other Provisions, as plentiful and cheap as in *Europe*; and the Ships are easily supply'd with Water, which they fill in their Boats at the Bridge, being convey'd thither in Wooden Pipes from the Bottom of the Hill. The *English* are not permitted to go up into the Country, but have all the Liberty they can desire of the Town and Gardens about it; as also to walk in the Company's Garden, which is large, at the upper End of the Town, with several fine Walks of Trees of sundry Sorts, one of them mostly of very thriving Oaks, at least half a Mile in Length, and very full of Acorns, besides others of all Variety of Fruit. It is divided into several Parts by Hedges of Rosemary and Lawrel, of a great Height, and

The Town.

Gardens.

and curiously cut; one Part is a Nursery of young Trees, another a Vineyard, another an Orchard, another affords all Sorts of Roots and Herbs for the Kitchin, and so others of divers Kinds. It is water'd by a pleasant Brook, running from the Foot of the neighbouring Mountain along the Hedges All Parts are thoroughly improv'd, to answer what it is design'd for, that is, to afford all Variety of Refreshments. In a little House in this Garden, are the Skins of several Beasts stuff'd, well deserving to be taken Notice of, there being a Lion and Lioness, a Mountain-Cow, a strip'd Ass, a Rhinoceros, an Elk, various Sorts of Antelopes, and other uncommon Creatures. The Houses are low built, tho' after the *European* Manner; but they are forc'd to thatch them on Account of the violent Winds.

Houses.

The Country this Town is seated in, some will have to be that commonly call'd *Monomotapa*, but it is more properly the Land of the *Cafres*, which Name is given to a great Part of that *Southern* Coast of *Africk*, both on the *East* and *West* Sides, and round the Point. It is very full of Mountains in this Part we are now speaking of, and consequently there are vast Numbers of wild Beasts, mighty Swarms of Birds of several Sorts, and no less Variety of Insects and venomous Creatures, too tedious for me to take Notice of. Neither will I pretend to be so good a Botanist, as to treat of the Plants, or other Things out of my Province, or which I could not have Leisure to enquire into. What I could learn concerning the Natives, during my Stay there, may perhaps not be disagreeable. These People are call'd *Hottentots*, their Skin is naturally swarthy, but by greasing, and other dawbing,

The Country.

Beasts, Birds, &c

ing, they become all over of a dirty black, their Limbs indifferently well made, their Heads long, their Hair woolly, and generally clogg'd with Sheeps Dung, or other Filth, their Noses flat, because crush'd down in their Infancy. As to Disposition, they are crafty and perfidious, and given to most Vices, and very fond of living idle, indulging themselves in Lust and Debauchery. They wear raw Guts about their Necks and Legs, which look like Puddings, having much Ordure in them, and these serve both for Food and Ornament, being eaten by them raw as they are, and filthy, when tender'd by being almost rotten. Others wear greasy Thongs of stinking Leather. The Women and better Sort of Men, wear generally a nasty untann'd Hide, or a Sheep's Skin, or that of any other Beast, the hairy Side next them, and hanging about their Shoulders, which being rubb'd over with Grease and Ordure, they stink abominably. The Women wind Sheeps Guts about their Legs, which at a Distance look like Rolls of Tabacco, and about their Middle have a Skin, with a Flap hanging down before, sometimes with Beads about it, which serves to hide what Modesty forbids to be seen. However, for a *Dutch* Doubleke they will shew all to the waggish Sailors, that ask them. They are all great Admirers of Tabacco and strong Waters. Men, Women, Children, and any other Creatures they have, lie together in little low Huts, cover'd with Skins, and stinking intollerably. They are the most filthy beastly People of any yet discover'd, and harden'd in their Brutality; for those who have convers'd with them, say, it is impossible to reclaim them. A Gentleman,

man residing at the Cape, told me, that sometimes, when any of them live to a decripid old Age, they shut them up in a Hut, and with them a Sheep, or some other Provisions, which, when spent, they bring no more, but leave them to famish to Death. Amidst this Beastliness, they are not altogether ignorant of the Being of a Deity, whom they call the Great Captain, and say he is angry, when there happens any Storm of Thunder and Lightning. They pay some worship to the Sun and Moon, and when she shines at Night, dance and revel, in Honour of her. Every Man has as many Wives as he can maintain, which are debauch'd by others before they marry them. On the Wedding-Day, the Bridegroom tells the Woman's Kindred, he can maintain her, and is past a Boy, in Testimony whereof he pisses upon one of them. Then they make merry, and rejoyce all together. Most of these Men have but one serviceable Testicle, the other being crush'd in their Infancy, which they pretend enables them to please more Women, as I was told in *Dutch* by an *Hottentote*. *Religion.*

They have nothing to trade with, but Cattel, which they exchange with the *Dutch* for strong Waters, Tabacco, and any Sort of Beads, the Manner being to shew how much the latter will give for an Ox, or a Sheep, and adding to it, 'till the Owner is satisfy'd. Every Village is under the Government of a Captain, whom all the Inhabitants honour and obey, insomuch that they dare not marry without his Leave, his Will being all the Law they are acquainted with. As for Weapons in War, they use Bows and Arrows, as also Darts, at casting which they are extraordina- *Weapons.*

ry dexterous, and often use to poison the Points, which generally proves mortal; and they wear a Sort of Leather Jackets by way of Armour. They are bold enough among themselves, their Weapons being equal; but care not to engage where there are Fire-Arms, unless it be in very wet Weather, when they think they may be out of Order. The only Handicrafts among them, are a Sort of Taylors, who sew the Skins they wear together, and Barbers, who trim their Heads in several Shapes. Nature has taught them some Cures for Wounds and Diseases. Their Language is extremely uncooth to our Ears, as all their Course of Life is disagreeable.

During our Stay at this Place, we employ'd the Time in Victualling and Watering, and making all other Dispositions for our Return Home. The *Duke*, commanded by Capt. *Woodes Rogers*, having been leaky ever since our being at *California*, where we had extraordinary good Weather, and generally smooth Water, and her Keel not being heav'd out, when we careen'd at the Island *Horn*, in *Batavia* Road, the Leak still continu'd, so that with the Bonnet on, she made nine Inches Water a Glass, and with the Bonnet off, we did suppose she would make twice as much, the Carpenters believing it was a Trennel-hole near the Step of the Fore-mast, by Reason it was constant. Notwithstanding all that could be done, this Leak continu'd 'till a few Days after we left the Cape, and then stopp'd of it self. The Committee empower'd Capt. *Rogers* and Capt. *Courtney* to carry one hundred Weight of Plate ashore, 60 Ounces of wrought Gold, and all the coin'd Gold and Silver in both their Ships, and, in Conjunction with Capt.

The Duke *leaky.*

Goods sold for Provisions.

Capt. *Dover* and my self, to purchase Necessaries wanting for them all, and to sell what Goods were proper for that Place, rather than exchange more Gold and Silver. Capt. *Courtney* was particularly directed to sell some Goods, and half a Dozen Blacks from Aboard the *Dutchess* and *Batchelor*.

Capt. *Opey*, Commander of an *East-India* Man, sailing from the Cape before us, we sent by him a Letter to our Owners, and a Copy of the same a few Days after by a *Dane*, which was as follows.

To Alderman *Batchelor* and Company, Owners of the *Duke* and *Dutchess* Frigats.

Gentlemen,

THIS *is to acquaint you with our safe Arrival at the Cape of* Good Hope, *which was on the* 29th *of* December 1710, *with our Prize the* Manila *Ship, call'd* Nuestro Senora de la Encarnacion y Defengano, *commanded by Monsieur* John Pichberty, *and now by us nam'd the* Batchelor *Frigat, mounted with* 20 *great Guns, and* 20 *Brass Pedrero's, and mann'd with* 116 *Men, a sound Ship. Each of our Ships is mann'd with* 120 *Men, and in Company with three* East-India *Ships, and expect three Sail more of* English *every Day. The* Dutch *from* Batavia, *being* 12 *Sail of stout Ships, are expected here every Hour, and* 6 *Sail more from* Ceylon; *which Fleet we have resolv'd in Council to accompany to* Holland, *unless we are assur'd of a Peace, or happen to meet with an* English *Convoy in crossing our Latitude. Our Ships are fitted with all Necessaries, and only wait for the Fleet, which we hope will sail by the last of* March; *hoping God will direct us, that we may come with Speech and Safety to*

Lett r to the Owners.

your

your *felves, and the reft of our Friends, to whom we render all due Refpects, and remain,*

Gentlemen,

Your moft humble, and moft obedient Servants,

Thomas Dover, *Prefident,* Woodes Rogers, Steph. Courtney, Ed. Cooke, Wil. Dampier. Robert Fry, Will. Stretton, Charles Pope, John Connely, Robert Glendal, John Ballet.

The *Dutch* Governor, and all other Perfons in Command at the Cape, treated us very civilly, and gave full Liberty to fupply our Ships with all that Place would afford.

February the 22d, the *Dutch* Fleet arriv'd from *Batavia,* being ten Sail, and two more in Sight, with three Flags. We faluted the Admiral, at his coming in, with feven Guns, as did the Fort with all or moft of theirs; and fo the *Dutch* Ships, for which the Flag return'd Thanks. When the Boats went with him afhore, he was faluted again, and fo at the Fort with their Cannon and feveral Volleys of fmall Shot. There were at this Time in the Harbour 17 *Dutch* Ships, 12 of them homeward bound, and 6 *Englifh.* During our Stay here, we bury'd on the 12th of *February* Mr. *Carlton Vanbrug,* the Owner's chief Agent, which was done as he defir'd of me, being one of his Executors, in a decent Manner in the Church-yard, moft of the *Englifh* Gentlemen there attending the Corps to the Church, the Ships firing Guns every half Minute, as is cuftomary on fuch Occafions. Befides him, dy'd Mr. *Appleby,* Mate to the *Duke,* and fome others.

After

Round the World. 75

After the 22d of *February*, we were taken up in killing Cattel for our Provision, and furnishing other Necessaries for our Voyage into *Europe*. The *Batavia Dutch* Fleet was soon follow'd by four Ships from *Ceylon*. Six came from that Island, but near *Madagascar* met with such a violent Storm, that some were forc'd to cut away their Main Masts, and throw over several of their Lee Guns, having much Water at the same Time in their Holds. The four came into the Cape much damag'd, and believ'd the other two had founder'd at Sea. Several *English India* Men arriv'd also in the Road, and some *Dutch*, these last from *Europe*, as also a *Portuguese* Ship from *Rio de Janeiro* in *Brazil*, and bound for *Mozambique*, on the *East* Coast of *Africk*, to take in Slaves. This is all that happen'd worth observing during our Stay at the Cape, we will now proceed on our Voyage.

CHAP. VII.

The Dutch *Admiral's sailing Orders.*

I Have inserted this Chapter of sailing Orders and Signals, believing they will be acceptable to all curious Persons, who desire to understand in what Manner Ships are able to converse with, and understand one another at Sea.

For

A Voyage *to the* South Sea, *and*
For the Commanders of the homeward-bound Ships, in their Voyage from the Cape of Good Hope, under the Flag and Command of Peter de Voss, Admiral of the Fleet bound for Holland.

The *Northbeek*, Admiral,
Herfetelede Leevice, Vice-Admiral,
Waffanaar, Rear-Admiral.

Dutch.	English:
Barnevelt,	Duke Frigat,
Stantvaftigheiit,	Dutchefs,
Meynden,	Batchelor,
Limburgh,	Donegal,
Avanturier,	Loyal Blifs,
Oofterfteyn,	Blenheim,
D'hemert,	Loyal Cook,
Arion,	Charlton.
Gamron,	
Corfloot,	
Hetraad huys van Middleburg,	
Bentveld,	
Enbeverwiick.	

Signals by Day.

TO know the *Admiral*, Vice-Admiral, and Rear-Admiral, they to wear their respective Flags.

To unmoor, to loofe the Fore-Top-Sail, and fire a Gun.

To weigh, hoift his blue Enfign, and fire a Gun.

To tack, he'll hoift his blue Jack at his Mizen-Peak, and fire one Gun. The Sternmoft and Leewardmoft Ships to tack firft.

To bear away, to wear, he'll hoift his red Pendant on his Enfign-Staff, and fire one Gun.

The

The Sternmoſt and Leewardmoſt Ships to wear firſt.

To miſs Stays. In caſe two Ships ſtand athwart each other, the Ship under Command in Steeridge and Way, to bear away firſt.

When you ſail by a Wind, or lie by, to bear away, he ſhall hoiſt his Enſign at his Mizen Top-maſt Head, and fire one Gun.

He that wou'd ſpeak with the Admiral, muſt hoiſt his Jack in his Place, and fire 'till the Admiral anſwers him with a Gun.

In caſe of loſing Company, and meeting again, the Leewardmoſt Ship to lower his Fore-Top-Sail, clew up his Fore-Sail, and fire a Gun; and the Weathermoſt to anſwer, by lowering his Main Top-Sail, haling up his Lee clue Garnet, and firing one Gun: But in caſe of blowing hard, and the Top-Sail not being out, then the Leewardmoſt to clew up his Mizen, hale him out again, and fire a Gun; if the Mizen is not out, then to hale him out, and make the Sign, and fire two Guns, and the Windwardmoſt to anſwer with his Main Stay-Sail, to hoiſt and lower it three times; but if he has not his Main Stay-Sail, he is to hoiſt it, and make the Signal, and fire two Guns. After the Signals are made and anſwer'd, he may come into the Fleet.

In caſe of ſeeing any Danger. If on the Starboard-ſide, to hoiſt a red Pendant at the Fore Top-maſt Head; if on the Larboard-ſide, a *Dutch* Pendant at the ſame Maſt, and fire one Gun; and every Ship that ſees it, is to fire a Gun, to ſhew they know it.

In caſe of ſeeing Land. He who ſees it, is to hoiſt his Jack and Enſign, and fire one Gun, and not to lower 'till the Admiral anſwers with his.

No

A Voyage *to the* South Sea, *and*

No Ship to go a-head of the Admiral. If any does by Day, he shall pay ten Dollars; if by Night 20 Dollars, to be on Account of the Mate who has the Watch.

In case of Need, to set up the standing Rigging. He who needs it, to hoist a red Flag at his Fore Top-mast Head; and if he sails well, may get a-head of the Admiral, that they may not lose Time; if he sails not well, he must make the same Signal, and the Fleet will tarry.

In case a Ship should run on Shoals, or Rocks, he must hoist out his Jack half Way upon his Ensign-staff, and get off as soon as he can, 'till other Ships answer; and if he cannot get off, all Ships are to send their Boats to his Assistance.

In case of calling a Consultation, the Admiral will hoist a white Flag on his Ensign-staff, and fire one Gun, that every Commander may repair Aboard him, and bring with him his Latitude and Longitude.

In case the Admiral would speak with the Vice-Admiral, Rear-Admiral, or any other Commander. For the Vice-Admiral, he will hoist a *Dutch* Pendant on the Ensign-staff, and fire one Gun. For the Rear-Admiral, a Pendant on the Mizen-Peak, and fire one Gun. For the *Standvastigheiit*, a *Dutch* Jack on his Mizen Top-mast Head, and fire one Gun. For the Commander of the *Limburg*, a white Flag on the Mizen Top-mast Head, and fire one Gun. For the Commander of the *Tamert*, a blue Flag on the Mizen Top-mast Head, and fire a Gun. For the Commander of the *Barnevelt*, a blue Pendant on the Mizen Top-mast Head, and fire one Gun. For the Commander of the *Oostersteyn*, a white Flag on the Mizen-Peak,

and

and one Gun. For the Commander of the *Meynden*, a *Dutch* Pendant at the Larboard Yard-Arm, and one Gun. For the Commander of the *Avanturier*, a red Pendant at the Mizen Top-maſt Head, and one Gun. For the Commander of the *Arion*, a red Jack at the Mizen Top-maſt Head, and one Gun. For the Commander of the *Raadhuys Van Middleburgh*, a *Dutch* Pendant on the Starboard Main Yard-Arm, and one Gun. For the Commander of the *Gamron*, a *Dutch* Pendant at the Starboard Crochet Yard-Arm, and one Gun. For the Commander of the *Corſloot*, a red Pendant at the Starboard Crochet Yard-Arm, and one Gun. For the Commander of the *Bentveld*, a *Dutch* Pendant at the Larboard Crochet Yard-Arm, and one Gun. For the Fiſcal of the *Meynden*, a white Pendant at the Starboard Main Yard-Arm, and one Gun.

When the Admiral will have any of the Engliſh Commanders come Aboard in particular. For the *Duke Frigat*, an *Engliſh* Jack at his Enſign-Staff, and one Gun. For the *Dutcheſs*, a King's Jack at the Mizen Peak, and one Gun. For the *Batchelor*, a King's Jack at the Mizen-Top-Maſt Head, and one Gun. For the *Donagal*, a King's Jack at his Larboard Crochet Yard-Arm, and one Gun. For the *Blenheim*, a King's Jack at his Mizen Top-Maſt Larboard Back-ſtay, and one Gun. For the *Loyal Bliſs*, a King's Jack at his Starboard Crochet Yard-Arm, and one Gun. For the *Loyal Cook*, a King's Jack at his Starboard Mizen-Top-maſt Back-ſtay, and one Gun. For the *Charlton*, a King's Jack at his Main Top-maſt Larboard Back-ſtay, and one Gun. For the *King William*, a King's Jack at his Starboard Main Top-maſt Back-ſtay, and one Gun.

When he would have all the English Commanders come Aboard him, he will then hoiſt his Double Prince Flag at his Mizen Top-Maſt Head, and fire one Gun.

To alter the Compaſs, he will hoiſt his red Flag at his Mizen-Peak, and fire a Gun.

To alter the Courſe, he will hoiſt a *Dutch* Flag at his Mizen-Peak, and fire a Gun.

In caſe of ſeeing a ſtrange Ship, he who firſt ſees it, to fire a Gun, then hoiſt and lower his Enſign, as often as he ſees Ships.

Signals for Chaſing. The Admiral will make the ſame Signal as he does for the Commanders to come Aboard, and hoiſt an Enſign on that Side he ſees the Ship, in his Mizen-Shroud, and fire a Gun.

To forbear Chaſing, he will hoiſt a blue Flag at his Fore Top-Maſt Head, and fire a Gun.

To anchor, he will hoiſt double the Prince Flag, and fire two Guns.

To moor, he will hoiſt his Mizen Top-Sail, with the Sheets clew'd up, and fire one Gun.

To cut or ſlip, hoiſt his blue Jack on his Enſign-ſtaff, and two Guns, and looſe both his Top-ſails.

In caſe of Fire, the Ship on fire to fire five Guns, and clew up his Low-ſails; and if at Anchor, to hoiſt a red Flag on his Enſign-ſtaff, and every Ship to anſwer with a Gun, and ſend their Boats to his Aſſiſtance.

In ſpringing a Leak, to hoiſt a Bonnet or Piece of Canvas at the Fore Top-Maſt Head, and fire Guns as quick as he can, and every Ship that ſees it, to fire one Gun, and come to his Aſſiſtance as ſoon as may be.

Loſing a Maſt, or other Accident, that you cannot keep Company with the Fleet, make a Waif with the Enſign, at the Fore Top-Maſt Head, and

JOURNAL-TABLE of our *Voyage*, in the Ship *Dutchess*, from the We... the Bounds of *Asia*, to the Cape of *Good Hope* in *Africa*. *Java* Head being in the Latitude of 7 D... Min. *East*; the Cape in the Latitude of 34 Deg. 15 Min. *South*, and Longitude 17 Deg. *East*. Perform...

	Course cor-rected.	Dist. fail'd.	Northings in Miles and Tenths.		Southings in Miles and Tenths.		Eastings in Miles and Tenths.		Westings in Miles and Tenths.		Latitude per Observation		Latitude per Estimation.		Longi-tude.		Meridian Distance.		Wind.
		Miles.	Miles.	Tenths.	Miles.	Tenths.	Miles.	Tenths.	Miles.	Tenths.	Deg.	Min.	D g.	Min.	D.	M.	D.	M.	
26,27	S.W.	112	0	0	85	0	0	0	85	0	0	0	8	30	1	26	1	25	S E.
28	S.W.by W.	50	0	0	50	0	0	0	75	0	9	0	9	0	2	41	2	40	E. S. E.
29	S.W.	121	0	0	85	0	0	0	85	0	10	25	10	25	4	9	4	5	S. E.
30	S.W.by W.	120	0	0	66	7	0	0	99	7	11	32	11	32	5	46	5	43	S. E. by E
Nov.1	W.S.W.½S.	230	0	0	109	0	0	0	216	0	13	20	13	20	9	35	9	29	S. E. by E
2,3	W.by S.¼ S.	235	0	0	70	2	0	0	221	4	14	39	14	39	12	20	12	10	E. S. E.
4,5	W.by S.¼ S.	200	0	0	58	0	0	0	191	4	0	0	15	37	16	57	16	31	E.
6,7	W.½ S.	150	0	0	14	0	0	0	147	0	0	0	15	51	19	32	19	0	Variable
9,10	W.	181	0	0	0	0	0	0	181	0	0	0	15	51	22	30	22	0	Variable
11,12	W.½ S.	154	0	0	22	0	0	0	152	0	16	13	16	13	25	17	24	34	S. N. E.
13,14	W.by S.	245	0	0	47	0	0	0	224	0	17	0	17	0	29	8	28	32	E.N.E.&c
15,16	W.S.W.	270	0	0	103	0	0	0	249	0	0	0	18	43	33	28	33	0	E. S. E.
17,18	W.S.W.½ S.	210	0	0	99	0	0	0	185	2	0	0	20	22	36	45	36	0	S. E.
19,20	W.½ S.	230	0	0	22	0	0	0	228	0	0	0	20	44	40	42	39	48	S. E.
21,22	W.S.W.	220	0	0	84	1	0	0	203	0	0	0	22	8	44	28	43	11	S. by W.
23,24	W.	230	0	0	107	1	0	0	208	0	0	0	24	27	48	6	46	39	S. E.
25,26	W.S.W.	170	0	0	103	0	0	0	246	0	0	0	0	0	52	35	50	48	E.
27,28	W.by N.	125	24	4	0	0	0	0	122	6	25	46	25	46	54	47	52	40	S. E. to S.
29,30	S.W. by W.¼ S	74	0	0	44	0	0	0	59	0	26	30	26	30	55	57	53	30	S. to E. E
b. 1,2	S.W.½ W.	120	0	0	76	0	0	0	83	0	0	0	27	46	57	26	55	2	S. E. to N
3,4	W.S.W.	271	0	0	104	0	0	0	251	0	29	30	29	30	62	2	59	23	S. E.
5,6	W.by S.½ S.	148	0	0	42	0	0	0	142	0	30	12	30	12	64	40	61	43	E.S.E. to S
7,8	W.S.W.¼ S.	160	0	0	75	0	0	0	141	0	31	27	31	27	67	25	64	4	S.S.E. to
9,10	W.by S.¼ S.	100	0	0	29	0	0	0	95	7	0	0	31	30	69	17	65	32	E.
11,12	W.by S.¼ S.	100	0	0	29	0	0	0	93	7	32	30	32	30	71	8	67	14	E. to S.
13,14	W.½ S.	160	0	0	15	0	0	0	158	0	0	0	32	45	74	26	69	52	S. W. to
15,16	N.	101	3	0	0	0	0	0	100	0	32	30	32	30	76	16	71	12	N. E. to
17,18	W.by S.¼ S.	56	0	0	16	0	0	0	53	0	32	46	32	46	77	16	72	25	S. E.
19,20	S.S.W.½ W.	100	0	0	68	0	0	0	47	0	34	14	34	14	78	16	73	12	E. to S.W.
21,22	W.by S.	151	0	0	30	0	0	0	149	0	34	44	34	44	81	25	73	43	E.
23,24	N.W.	20	14	0	0	0	0	0	14	0	34	30	34	30	81	25	75	55	E.S.E to W
25,26	W.	30	0	0	0	0	0	0	30	0	34	30	34	30	82	9	76	25	S. b W. to
27,28	W.S.W.½ S.	70	0	0	33	0	0	0	61	0	35	3	34	3	83	27	77	20	S. W. to

and fire two Guns; then the Ship that is nearest, is to go to the Assistance of the Distress'd, 'till the Admiral stays for him; and when he is assisted, he is to take in his Waif, and fire one Gun.

To scud under a Fore-sail, the Admiral will hoist a red Jack on his Ensign-staff, and fire three Guns.

In case of bringing to, to sound, he is first to fire five Guns.

In case of Ground, he is to hoist his Jack at his Ensign-staff, and one Gun.

Signals by Night.

1. *Lights by Night*. The Admiral constantly carries a Light at his Main-Top, and one at his Stern. The Vice-Admiral and Rear-Admiral, one at their Stern. When the Admiral puts out two Lights in his Stern, then the other Admirals are to put out two, and every other Ship one.

2. *To anchor*, the Admiral will, besides his usual Lights, put one at the Main Shrouds, and one at his Fore-shrouds, of equal Height, and fire two Guns, and then every Ship to put out two Lights of an equal Height, at the same Place, and when let go their Anchor, fire two Guns.

3. *To weigh*, the Admiral will, besides his usual Lights, put out two Lights at his Main Top-Mast Shrouds, of equal Height, and fire one Gun; and then every Ship to put out two Lights at his Stern, to shew he sees it. When the Admiral is under Sail, he will fire one Gun, and take in his Light-Signals; and every Ship, when under Sail, to take in one Light, and fire one Gun.

4. *To tack*, the Admiral will hoist a Light at his Ensign-staff, and fire a Gun; and every Ship to put out a Light, to shew he sees the Signal; and before he tacks, to hoist the same Light at his Ensign-staff; and when tack'd, fire one Gun, the Sternmost Ship to tack first.

5. *To wear*, the Admiral will hoist a Light at his Mizen-Peak, and fire two Guns, and every Ship to answer his Signal, by shewing a Light at his Mizen-Peak. The Sternmost and Leewardmost Ships to wear first; and every Ship when wore, to fire two Guns.

6. *To hand the Top-sail*, the Admiral will put out three Lights on his Main Shrouds, of an equal Height, and fire one Gun, and every Ship to answer by the same Signal.

7. *To lye by*, the Admiral will put out four Lights in his Fore-shrouds, of equal Height, and fire four Guns one after another, and every Ship to answer with two Lights in their Fore-shrouds, and fire two Guns, and not to take in their Lights 'till the Admiral has.

8. *To make Sail*, the Admiral will put out two Lights, one above another, in his Fore-shrouds, and fire three Guns; and every Ship, when the Fore-sail is set, to answer with one Gun.

9. *To bring to, with the Fore-sail to the Mast*, the Admiral will put out a Light in his Main Shrouds, and his Main Top-Mast Shrouds, and fire two Guns, and every Ship to answer by the same Signal of Lights, firing two Guns.

10. *To furl the Fore Top-sail*, the Admiral will put a Light in the Fore Top-Mast Shrouds, and fire two Guns, and every Ship to answer by

by the same Signal of Lights, and firing two Guns.

11. *To know one another by Night*, in case any Ship loses the Fleet, and comes in again by Night, he is to hail from whence your Ship, and the other to answer from *Guada*; then the other from *Groe, God be with us*; and as you have spoke, *God be with us*, to fire one Gun to Windward, the same to be answer'd by the Ship that is hail'd, with two Guns to Leeward.

12. *In case of seeing strange Ships by Night*, any one seeing a strange Ship, is to put out a Light at his Main Top-Mast Head, and another at his Mizen-Peak, and lower his Light at his Mizen-Peak, as often as he sees Ships, and fire sometimes a Gun, and make false Fires, 'till the Admiral puts a Light at his Main Top-Mast Head.

13. *In case of not being able to keep Company with the Fleet*, the Ship that is not able to keep Company by Night, to put out three Lights, one at the Sprit-sail Top-Mast Head, and the others at each Sprit-sail Top-sail Yard-Arm, and fire two Guns; the Admiral, and all other Ships to answer with one Gun, and every Ship to keep such Sail, as he thinks the other can keep Company with him, but in case the first judges himself capable to make more Sail, then he is to put out a Light at his Fore Top-Mast Head, and fire two Guns; then the Admiral and every Ship to answer with one Gun.

14. *In case of Fire, or dangerous Leak*, to fire seven Guns, and put out as many Lights as he can, and not to take them in 'till the Fire is out, or the Leak stopp'd; in the mean Time, every Ship to send their utmost Assistance.

15. *To alter their Courſe*, the Admiral will hoiſt three Lights one above another, at his Mizen-Peak, and fire one Gun, and every Ship to anſwer by the ſame Signal.

16. *To bear away*, the Admiral will put out a Light at each Main Yard-Arm, and fire three Guns.

17. *To lye by in bad Weather*, when you bring to with Larboard Tacks on Board, the Admiral to put out three Lights of equal Height at his Fore-ſhrouds, and fire five Guns; and when he lyes by, his Starboard Tacks aboard, the Admiral will fire ſeven Guns, and the ſame Lights as before, or the firſt Signal, and every Ship to anſwer with one Gun, or the latter Signal with three Guns.

18. No Ship to bring to, 'till the Admiral has made the aforemention'd Signal, N. 17.

19. *To ſound in the Night*, any Ship deſiring to ſound, muſt firſt put three Lights aſtern, and fire one Gun.

20. Any Ship finding Ground, is to put four Lights one above another at his Mizen-Peak, and fire one Gun.

21. Any one ſtriking Ground, without bringing to, is to fire one Gun, and put out two Lights at his Mizen-Peak one above another, and to hoiſt and lower them two Hours, or 'till the Admiral anſwers with one Gun.

22. Any Ship ſeeing Land or Shoals, is to put out three Lights one above another at his Enſign-ſtaff, and two at his Mizen-Peak, and fire three Guns, and then go about; then every Ship to tack, and fire one Gun. But in caſe the firſt cannot get off, to make the ſame Signal the Admiral does for Anchoring, and every one to make the beſt of his Way to anchor or get off.

23. Any Ship running on a Shoal, is to put

put a Light at the three Top-Maft Heads, and fire Gun after Gun, 'till every Ship anfwers with two, and every one to fend Affiftance.

Signals in a Fog.

1. ON a Mift firft rifing, the Admiral will shorten Sail 'till the worft Sailors come up with him, and keep the fame Sail during the Mift, and fire two Guns every half Hour, and every one of the Ships to anfwer one after another, with one Gun, and now and then to fire a Mufket, beating a Drum continually, and on the Mift rifing, to get fuch a Berth as to keep clear of one another.

2. *To tack*, the Admiral will fire fix Guns; the Sternmoft to tack firft, and when tack'd, to fire fix Guns.

3. *To lye by*, when the Admiral brings to with his Larboard-Tacks, he will fire three Guns, and with the Starboard-Tacks five Guns; and when every Ship on his Larboard-Tack is brought to, muft anfwer with three Guns, and with the Starboard with two.

4. *To go the fame Courfe again*, the Admiral will make the fame Signal, as he does for failing, firing four Guns one after another, and muft be anfwer'd by every Ship with four Guns, as foon as he has made Sail.

5. *To anchor*, the Admiral is to fire feven Guns, and every Ship anchor'd, anfwer with five Guns; and when the Admiral has been at Anchor half an Hour, he will fire feven Guns, in cafe any of the Sternmoft Ships heard not the firft.

6. *To weigh*, the Admiral will fire four Guns, one after another, and is to be anfwer'd by every Ship, when under Sail, by four Guns.

7. *In case of seeing Shoals or Rocks,* he who sees them, must make the same Signal to go about, as the Admiral does to tack in a Mist; but if he thinks better to anchor, he is to make the same Signal as the Admiral does to anchor in a Mist. In case of being fast ashore, to keep firing Guns, that the other Ships may tack, or come to an Anchor, and send Assistance.

8. *In case of not being able to keep Company with the Fleet,* such Ship is to fire eight Guns, and then the Admiral will shorten Sail, or lie by 'till he comes up.

9. *To alter the Course,* the Admiral will fire nine Guns, and is to be answer'd with nine Guns by every Ship.

10. *In case of striking Ground.* Any Ship striking Ground, and not imagining any Danger, is to fire eleven Guns, and keep on.

11. All the Signal-Guns mention'd to be fir'd in misty Weather, or by Night, are to be fir'd on one Side.

How Ships are to place themselves at Sea.

Barnevelt, Noorbeck, Admiral,
   Debemert, Den Avanturier must keep to
Dutchess, Windward, and close by
   Donegal, him, d'Herfteldeleevice, V. Ad.
Meynden, d'Arion,
   de Stantvastegheyt, Duke Frigat,
Bentveld, Loyal Cook,
   Charlton, Oostersteyn,
Blenheim, Limburgh,
   King William, Traadhuys,
Bsvenwyck, Loyal Bliss,
               Batchelor,
            Hersloot,
              Gamron.
Wassunaar, Rear-Admiral.

Order'd

Order'd by the Admiral, that all the foregoing Orders and Signals be strictly observ'd by every Officer, that has Charge of the Watch, on board every Ship.

Signals for the Ships to draw in a Line.

IN *case of seeing strange Ships*, the Admiral will hoist a red Flag at his Ensign-staff; every Ship to keep in the Line of Battel, sailing by the Wind, and afore the Wind; and when the Admiral desires to have Ships in a Line of Battel, he will hoist his *Dutch* Jack and Pendant at his Mizen-Peak, and fire two Guns. Then every Ship to come before the Wind in a Line of Battel, as is after laid down.

Wassanaar, Barnevelt, Dutchess, d'Hemert, King William, Bentveld, Charlton, Donegal, Beverwyck, Blenheim, Meynden, Noorbeck, Oostersteyn, Batchelor, Corsloot, Loyal Bliss, Gamron, Loyal Cook, Limburgh, Arion, Duke, Traadhuys, Hersteldeleevice.

When the Fleet sails in the above Order, and the Admiral will have them to hale upon a Wind, with the Starboard Tacks Aboard, in a Line, one after another, to attack the Enemy, he will hoist a blue Flag at his Mizen-Peak, and fire one Gun. Then the Vice-Admiral is to be the headmost Ship; and when his Larboard Tacks are Aboard, he will hoist a red Pendant under the aforesaid Flag, and fire two Guns Then the Rear-Admiral is to be the headmost Ship.

88   A Voyage to the South Sea, and

Line of Battel with the Starboard Tacks Aboard.	Line of Battel with the Larboard Tacks Aboard.
Vice-Adm. *Herſteldeleevice*, *Traadhuys van Middleburgh*, *Duke*, *Arion*, *Limburgh*, *Loyal Cook*, *Gamron*, *Loyal Bliſs*, *Corſloot*, *Batchelor*, *Ooſterſteyn*, Admiral *Noorbeck*, *Meynden*, *Blenheim*, *Beverwyck*, *Donegal*, *Stantwaſtegheyt*, *Charlton*, *Bentveld*, *King William*, *d'Hemert*, *Dutcheſs*, *Barnevelt*, Rear-Admiral *Waſſanaar*.	In this Caſe the Names of all the Ships are to be writ or read from the Bottom to the Top, the Rear-Admiral being a-head, and the reſt following upwards.

All the Ships in the aforeſaid Line, to keep as near one another as they can, and not give the Enemy Opportunity to break the Line.

When the Vice-Admiral is to tack firſt, the Admiral will hoiſt a red Pendant at his Fore Top-maſt Head, and fire one Gun.

When the Rear-Admiral is to tack firſt, the Admiral will hoiſt a blue Pendant at his
Mizen

Mizen Top-Maft Head, and fire one Gun.

When the Headmoſt Ship is to make more Sail, the Admiral will hoift a double Prince-Flag at his Fore Top-Maft Head, and fire two Guns.

When the Headmoſt Ship is to ſhorten Sail, the Admiral will hoift a broad *Dutch* Pendant at the Fore Top-fail Yard-Arm to Leeward, and fire one Gun.

To make two Lines, the Admiral will hoift a double Prince-Flag at his Mizen-Peak, and fire two Guns, failing with his Starboard-Tacks aboard; the Vice-Admiral with his Divifion to ſhorten Sail, 'till the Rear-Admiral comes up with him, and to make two Lines with the Larboard-Tacks aboard. The Rear-Admiral is to ſhorten Sail, 'till the Vice-Admiral comes up with his Divifion, to make two Lines as before.

In caſe any Ship ſhould be diſabled or over-match'd, any Ship being boarded, and finding the Enemy over-powering him, is to hoift a red Pendant at his Main Top-Maft Head. Then the Fly-Boats and fmall Ships are to affift him with what they are able, in caſe they are not engag'd themſelves.

When the Admiral hoifts a red, yellow, and white Flag, the Ships have Liberty to break the Line, and engage as they will.

To ſpeak with the Commanders, when the Admiral hoifts a double Prince's Flag at his Mizen-Peak, then all the Ships to run aftern one after another to the *Noorbeck*, to take his Orders.

Signal for the *Avanturier*, is a *Dutch* Pendant at his Fore Yard-Arm.

When the Admiral defires all Boats to come aboard arm'd, he will hoift a red Jack at his

Mizen

Mizen Top-Maſt Head, and fire one Gun, and every one to repair to him as faſt as he can.

Signals for Chaſing, as in the general Orders, viz. the Dutcheſs *Frigat, King* William, *and* Cline *Caper.*

IF you ſee a ſtrange Ship, and have a Signal for Chaſing, and come up with the Chaſe, you are to hail her, *From whence your Ship? How long have you been out? Whither are you bound? What is your Ship's and your Commander's Name?* And, *whether Peace or War with* France? If a Friend, command his Boat Aboard, with an Officer, and all the News-Papers he has; and if War, enquire where the *French* Fleet cruizes. If he refuſe to come Aboard with his Boat, you are to fire a Shot over him, and command him. But if he proves to be an Enemy, and too ſtrong for you, then you are to hoiſt a red Flag at the Fore Top-Maſt Head, and fire three Guns on the contrary Side the Ship or Ships are, which will be a Signal to the Flag. But if a Friend, and ſuch Weather that he cannot hoiſt his Boat out, he is to keep to Leeward of the Fleet, and bring to, and you to have a white Flag at your Main Top-Maſt Head, and to fire five Guns to Leeward, 'till the Admiral ſpeaks with him. But if he will not lye to at Command, then you are to hoiſt a red Flag at your Fore Top-Maſt Head, and fire three Guns on the contrary Side, as aforeſaid. When you come to the Latitude from 58 to 62, and happen to meet a *Swede, Dane,* or *Dutch* Ship, you are to command their Boats aboard, and oblige them to give Account, if

Round the World.

War or Peace, and to ſtay with the Fleet. If we meet with a Ship between this and the Tropick, and they ask, *From whence your Ship?* you are to anſwer, *from* Brazil, *and bound for* Liſbon, if not paſt *Liſbon*. But if paſt *Liſbon*, then ſay to *Amſterdam*.

CHAP. VIII.

Departure from the Cape of Good Hope; *Iſlands of St.* Helena, *and the* Aſcenſion, *in the* South ; Bora, *and* Shetland *Iſlands, in the* Northern *Sea* ; *Arrival in* Holland, *and what happen'd there* ; *Departure thence, and Arrival firſt at the* Downs, *and then in the* Thames.

*A*Pril 6. 1711. Having ſtor'd and fitted our Ships for ſo long a Run, as from the Cape of *Good Hope*, into *Europe*, we weigh'd in the Morning, and ſail'd with a ſmall Breeze of Wind at *E. S. E.* and by Noon came to an Anchor at *Penguin* Iſland, in nine Fathom Water, the *Eaſt* End of the Iſland bearing *South Weſterly.* At Three weigh'd again, being in all 25 Sail of *Engliſh* and *Dutch*, all good Ships, under the Command of Admiral *Peter de Vos.* At Six, the Cape of *Good Hope* bore *S.* by *W.* half *W.* diſtant ſeven Leagues. The 7th being haſy Weather, could not have an Obſervation, Wind at *S. S. E.* a ſmall Breeze, and fine Weather. I have here inſerted the Journal-Table from the Cape, 'till our meeting of the *Dutch*

Cruizers

Cruizers in our *Northern* Seas, wherein are the Latitudes, Longitudes, Currents, Variation, Weather, &c. during all that long Run from Day to Day.

The failing Part of this Voyage being contain'd in the Table, it only remains to add what else remarkable occurr'd during that Time, where once for all it may be said, that the *Batchelor* Prize not keeping up with the Fleet, we were oblig'd several Days to take her in Tow; yet sometimes alter'd without any Help, and made Way as well as the rest.

April 23. being St. *George*'s Day, and the Anniversary of her Majesty of *Great Britain*'s Coronation, the *Duke* and *Dutchess* saluted each other with several Chears, Drums beating, Trumpets sounding, and St. *George*'s Jack flying. In the Evening, gave all the Men Liquor, to drink her Majesty's Health.

St. Helena Island. *April* 3. 1711. in the Evening made the Island of St. *Helena*, at a good Distance. It lyes in 16 Degrees of *South* Latitude, and 22 Degrees Longitude *West* from the Cape of *Good Hope*, being about nine or ten Leagues in Length, not so much in Breadth, and above 300 Leagues from the Continent of *Africk*. Next the Sea it is almost every where encompass'd with high Rocks, which hinder the Approach, there being but one Place convenient for Landing; and within there is much Mountain, but most of it cover'd with wholsom Herbs and Plants; and the Valleys are so fruitful, that they produce whatsoever is brought from other Parts, and planted, in great Perfection. The *Portuguese* first discover'd it accidentally, as they were ranging down the Coast of *Africk*, to make their Way to the *East Indies*.

A JOURNAL-TABLE of our *Voyage*, from the Cape of *Good H*

Edward Cook. We weigh'd from *Penguin* Island the 6th of *April* in the Afternoon, in Company with

Months and Days.	Course corrected.	Dist. sail'd.	Northings in Miles and Tenths.		Southings in Miles and Tenths.		Eastings in Miles and Tenths.		Westings in Miles and Tenths.		Latitude per Observation.		Latitude per Estimation		Longitude.		Meridian Distance.	
		Miles	Miles.	Tenths.	Miles.	Tenths.	Miles.	Tenths.	Miles.	Tenths.	Deg.	Min.	Deg.	Min.	D.	M.	D.	M.
April 7, 8	N. 43 Deg. W.	192	142	0	0	0	0	0	130	0	31	53	31	53	2	28	2	10
9,10	N.N.W. ½ W.	140	113	0	0	0	0	0	65	0	29	50	29	10	3	41	3	15
11,12	N. W.	71	50	0	0	0	0	0	50	0	29	0	29	0	4	38	4	20
13,14	N. W. ½ N.	140	108	2	0	0	0	0	89	0	27	12	27	12	6	48	5	49
15,16	N. W. by N.	107	89	0	0	0	0	0	59	0	25	43	25	[illegible]	7	26	6	48
17,18	N. W. by W.	212	118	0	0	0	0	0	176	0	23	45	23	4	10	40	9	44
19,20	N. W.	116	82	0	0	0	0	0	82	0	22	33	22	3	12	12	11	6
21,22	N. W.	79	56	0	0	0	0	0	56	0	21	37	21	3	13	13	12	2
23,24	N. W. by W.	60	44	0	0	0	0	0	66	0	21	53	21	5	14	23	13	8
25,26	N. W. by N.	151	130	0	0	0	0	0	78	0	19	43	19	4	15	42	14	26
27,28	N. W. ½ W.	60	38	0	0	0	0	0	46	0	19	5	19	[illegible]	16	31	15	12
29,30	N. W.	208	147	1	0	0	0	0	147	1	16	38	16	3	19	7	17	39
May 1	☞										16	0	16	[illegible]	22	0	20	43
2,3	N. W. by W.	270	150	0	0	0	0	0	225	0	13	30	13	7	3	51	3	45
4,5	N. W.	235	166	1	0	0	0	0	166	1	10	44	10	4	6	42	6	31
6,7	N. W. ½ N.	230	178	0	0	0	0	0	140	0	7	46	7	[illegible]	9	41	8	57
8,9	W. 40 Deg. N.	177	150	0	0	0	0	0	180	0	0	0	5	0	3	1	3	0
10,11	N. W.	184	130	0	0	0	0	0	130	0	0	0	3	0	5	13	5	10
12,13	N. W.	185	130	0	0	0	0	0	130	0	0	30	0	0	7	23	7	20
14,15	N. W.	138	98	0	0	0	0	0	98	0	0	48	0	48	9	1	8	58
16,17	N. W.	180	127	0	0	0	0	0	127	0	3	5	3	5	11	8	11	5
18,19	N. W.	170	120	0	0	0	0	0	120	0	5	5	5	5	13	8	13	5
20,21	N. W.	51	36	0	0	0	0	0	36	0	5	41	5	41	13	44	13	41
22,23	N. W.	104	71	0	0	0	0	0	71	0	6	52	6	52	14	55	14	52
24,25	N. W.	100	70	1	0	0	0	0	70	1	8	2	8	2	16	5	16	2
26,27	N. W. ½ N.	179	138	4	0	0	0	0	113	5	10	0	10	0	18	0	17	56
28,29	N. W.	185	131	0	0	0	0	0	131	0	12	11	12	11	20	13	20	7
30,31	N. W. by N.	183	152	0	0	0	0	0	92	0	14	43	14	43	21	47	21	39
June 1,2	N. N. W.	180	166	3	0	0	0	0	69	0	17	29	17	29	23	0	22	48
3,4	N. N. W.	180	166	3	0	0	0	0	69	0	20	15	20	15	24	12	23	57
5,6	N. by W.	180	176	0	0	0	0	0	35	0	0	0	23	11	24	47	24	32
7,8	N. by W. ½ W.	150	154	0	0	0	0	0	43	0	25	34	25	35	25	25	25	15
10	N. by W.	80	78	5	0	0	0	0	15	0	20	53	26	53	25	51	25	31
11,12	N. by W.	110	108	0	0	0	0	0	21	0	28	41	28	41	26	17	25	52
13,14	N. by W.	140	137	3	0	0	0	0	27	3	0	0	30	58	26	44	26	19
15,16	N. N. W.	50	46	2	0	0	0	0	19	1	31	44	31	44	27	12	26	38
17,18	N. by W. ½ W.	80	76	6	0	0	0	0	23	2	0	0	33	0	27	35	27	1
19,20	N.	150	150	0	0	0	0	0	0	0	0	0	35	30	27	35	27	1
21,22	N. ½ E.	247	245	3	0	0	35	4	0	0	39	35	39	35	27	0	26	25
23,24	N. N. E.	300	277	0	0	0	115	0	0	0	44	12	44	12	25	21	23	40
25,26	N. N. E. ½ E.	293	258	0	0	0	138	0	0	0	48	30	48	32	21	9	20	12
27,28	N. E. by N.	170	141	3	0	0	94	5	0	0	0	0	50	51	19	14	18	14
29,30	N. E. ½ N.	170	131	4	0	0	108	0	0	0	53	2	53	2	16	11	16	2
July 1,2	N. E. ½ N.	230	178	0	0	0	46	0	0	0	0	0	56	0	12	9	13	32
3,4	N. E.	60	42	0	0	0	92	0	0	0	0	0	56	42	11	1	12	46
5,6	N. E. by N.	100	83	1	0	0	56	0	0	0	0	0	58	5	9	8	11	4
7,8	E. N. E.	170	67	1	0	0	157	1	0	0	0	0	59	2	4	14	8	27
9,10	E. N. E. ½ N.	181	85	0	0	0	169	0	0	0	0	0	60	7	1	19	5	40
11,12	E. ½ S.	286	0	0	42	0	0	0	0	0	0	0	59	5	11	1	0	58
13,14	E. S. E.	100	0	0	38	3	92	4	0	0	0	0	59	7	14	2	0	34
15,16	E.	100	0	0	0	0	100	0	0	0	0	0	59	6	17	10	2	14

☞ No, *May* 1. We made the Island of St. *Hellena*, whose Latitude and Longitude from the ab

Indies. According to their Custom, they left here some Goats and Swine, which increas'd so considerably before the Island was inhabited, that they have ever since afforded Refreshment to Ships that touch there. The *Dutch* first inhabited, and some Time after abandon'd it for the Cape of *Good Hope*, when the *English* possess'd themselves of it; but the *Dutch*, who had not thought it worth keeping before, then thought fit to turn them out; yet kept it not long, being some Time after expell'd by the *English*, who still keep it, having erected a good Fort for its future Security, and built a small Town near the little Bay, where Ships generally anchor in their Way to or from *India*, to water and refresh their Men. The Island producing Oranges, Lemmons, all Sorts of Greens and Roots, the Sailors find much Relief at it against the Scurvy. Besides, it affords Plenty of Black Cattel, Swine, Fowls, Ducks, Geese, and *Turkeys*, which the Inhabitants sell for any Sort of Cloathing, or other Necessaries they stand in need of. It would be delightful living there, were the Spot larger, nearer some Christian Country, or more resorted to; but the Confinement to so small a Place in the midst of a vast Ocean, so remote from Communication with the rest of the World, renders it uncomfortable. Did not our *India* Ships often touch at the Cape of *Good Hope*, the Inhabitants of St. *Helena* would fare something the better, all their Trade depending on those Ships, which being but few, the want of any of them becomes a considerable Damage. The Sea round about affords Variety of good Fish, which is another Help to those who have liv'd long upon salt Provisions. Sick Persons are often set ashore whilst the

the Ships ſtay; and if their Diſtemper be occaſion'd by the Sea, ſoon recover by Help of the wholſom Air of the Place, and the Variety of Refreſhments. It is a mighty Satisfaction to Sailors, as well as Paſſengers, to ſet their Feet aſhore, after ſo long a Paſſage as is from *Europe* thither, and before they proceed upon ſo great a Run as ſtill remains from thence to *India*. The *Dutch* have made the Cape what it is, from a wretched deſolate Place, only for the Relief of their *India* Ships; and were the *Engliſh* as induſtrious in promoting their own Intereſt, this Iſland might be conſiderably improv'd, and no leſs Advantages would accrue from it, the Situation being ſo convenient for that Trade, and the Soil producing all Things proper for Refreſhment.

For ſome Days paſt, found we had met with ſtrong Currents, ſetting us to the *N. W.* farther than we expected. *May* the firſt 1711. at Noon took our Departure from the Iſland St. *Helena*, bearing *Eaſt*, diſtant 11 Leagues.

May 7. 1711. after our Departure from St. *Helena*, to this Time, had freſh Breezes at *S. E.* as in the Journal Table, with ſtrong *S. E.* Currents. This Morning at Six made the Iſland of *Aſcenſion*, which ſhews to be bigger than St. *Helena*, which is not inhabited, has little freſh Water, therefore little reſorted to, unleſs ſometimes Ships touch there for Tortoiſes, whereof it has much Plenty. This Day at Noon it bore *E.* half *N.* diſtant 10 Leagues. I make its Latitude about 7 Deg. 40 Min. *South*; and Longitude from St. *Helena*, 9 Deg. 10 Min. *Weſt*. We ſtill found the Current ſet us to the *N. W.* and ſo continu'd

Aſcenſion Iſland.

inu'd to the 17th. Cross'd the Equinoctial the 13th.

May 23. were in about 7 Deg. of *North* Latitude, and since the 17th had small variable Breezes, with close, hasy, very hot Weather, and heavy Showers of Rain; but at this Time found little Current. The 22d the Admiral's Captain of the small Privateer had been Aboard, to bring our chasing Instructions. The *Dutchess* and *King William* Frigats, with the *Adventurer*, a *Dutch* Caper, were appointed for that Purpose; the Particulars whereof are among our sailing Orders from the Admiral.

June 8. two of our *Dutch* Sailors dy'd, and were decently bury'd, according to the Custom at Sea. Had very hot Weather for some Time before, and saw great Quantities of Gulph Weeds. About this Time all the Flag Ships struck their Flags, the Admiral hoisted a broad Penant, and all the other Ships theirs, which I take to have been done, that in case we should see any Ships, they might imagine us to be a Squadron of Men of War. The 12th, we whipp'd two *Dutch*-men severely for Mutiny and Quarrelling, and then put them into Irons. Had but little Wind for some Days, by Reason of our being in the calm Latitude, which we reckon from 25 to 28 Deg.

June 13. the Admiral made a Signal for each Ship to keep in the Line, the 14th black'd our Ship, the 15th in the Morning the Admiral made a Signal for all the *English* Commanders, who went Aboard and din'd with him, had a plentiful Entertainment, and were much caress'd. These Days small Gales from the *E. N. E.* and *N. E.* and very hot Weather.

June

June 22. were forc'd to caft off the *Batchelor*, which we would have tow'd, as was frequently done before; but could not then, by Reafon of the Squals coming ftronger; therefore the *Batchelor* had leave to keep a-head of the Fleet, and we to be near, for Fear of any Misfortune. The fqually Weather oblig'd us to go under our low Sails.

June 23. a *Dutch* Ship being in Diftrefs, fir'd feveral Guns, whereupon we fhorten'd Sail; but foon after the fame, firing another Gun, to fignify the Danger was over, we all made Sail again. The 24th endeavouring to tow the *Batchelor*, ftav'd two Cafk we had flung, and veer'd a-ftern, inftead of a Buoy for them; but they carrying fo much Sail, could not get hold of it. In the Afternoon the *Batchelor* hoifted out their Pinnace, and fent Aboard, to carry the End of our Cable from us to them, with a Coil of fmall Rope, to veer away upon Occafion; but through Careleffnefs of thofe Aboard the Ship, as well as in the Boat, fhe funk a-thwart the Hawfer; however, the Men all got Aboard. Then we turn'd to Windward, and took up moft Part of the Boat, the Ship having broke her Back in running over her. In the Evening the *King William* took the *Batchelor* in Tow, and caft her off again the next Morning. The 26th a Council of War was held Aboard the Flag. At this Time we again found a ftrong Current fetting us to the N. E. Latitude, Longitude, &c. as in the Journal-Table.

June 30. from the 26th had *Southerly* Winds and thick Weather, therefore kept firing Guns, as directed by our failing Orders, every half Hour. Continu'd the fame the 1ft and 2d of
July;

July; and being in above 58 Degrees of *North* Latitude, had little or no Night, but very thick cold Weather. Capt. *Courtney*, and many of our Men, were sick at this Time.

July 10. 1711. in the Morning, the *Donegal India-Man* having lost her Fore Top-Mast, made a Signal of Distress, whereupon the Fleet shorten'd Sail; but another Mast being got up, all sail'd again. *July* 12 in the Morning, reeft our Courses, then brought to, and lay by for the Fleet. Being at this Time in near 60 Degrees of *North* Latitude, had no Night, but cold drisling Weather; yet nothing comparable to the Cold going about Cape *Horn* in the same Latitude *Southward*. These Days saw several Gulls, and other Sea-Fowl. The 14th at Seven in the Morning, made the Island *Bora Bora*, bearing about *S.S E.* distant 8 Leagues, had then a moderate Gale at *S. W.* At Three in the Afternoon spoke with a *Danish* Vessel, bound for *Dublin*, who acquainted us there was still War between *France* and the Allies, as also between the *Danes* and *Swedes*, and that he had spoken with a *Dutch* Squadron of about 13 Sail of Men of War and Victuallers, that were cruizing in quest of us, near *Fair Island*, so that keeping between that and *Shetland*, they must needs see us. The 15th in the Morning, we saw those Ships to Windward of *Fair Island*. Having made the Signal, they bore down to us, and by Noon several of them had join'd us. Then all our Fleet saluted the Commodore, and he made the Signal for all the Commanders to go aboard him for their Sailing-Orders. Capt. *Courtney* and Capt. *Dover* went, and were very well entertain'd by the *Commodore*, who told them he would supply their Ships with Beer, or any

Bora Island.

Shetland.

H other

other Thing he had aboard, they paying for the fame. I reckon'd we were this Day at Noon in 59 Deg. 16 Min. Latitude *North per* Eſtimation. Longitude from the Iſland *Aſcenſion* to *Fair Iſland*, 17 Deg. 10 Min. *Eaſt*. Variation about 11 Deg. 30 Min. *Weſt*. We lay off *Shetland* two or three Days, for ſome of the Cruizers to join us; and having but little Wind, catch'd Ling, Cod, and other Sorts of Fiſh. The Inhabitants of *Shetland*, who are *North Britains*, went aboard our Ships in *Norway* Yauls, carrying freſh Proviſions, which they ſold at very reaſonable Rates, being a very poor People. It is not worth while to give any Account of this poor *Northern* Iſland, which can afford nothing remarkable; and indeed it is from my Purpoſe, being quite out of the Way as to our Voyage to the *South Sea*, or round the World. The Reaſon of our coming Home that Way, was for the avoiding of any ſtrong *French* Squadron which might be abroad, and we coming from ſuch remote and *Southern* Parts intirely ignorant of it. This Way *North* about has been often us'd, for the more Safety in Time of War with *France*, thoſe Seas being ſo remote from their Parts; and where they are unwilling to expoſe their Men of War, as well by Reaſon of the little Security there is from the Wind and Weather, as becauſe of the Danger of *Dutch* or *Engliſh* Fleets, which have nearer Places to retire to upon diſcovering any Danger; the neareſt the *French* can pretend to make for out of thoſe *Northern* Climes, being *Dunkirk*, well known to be none of the beſt, eſpecially for great Ships to go into boldly in Caſe of Danger, notwithſtanding the prodigious Expence

pence the King of *France* has been at for improving that Harbour.

July 19. at Noon I reckon'd the Head of *Shetland* bore *N. N. W.* diſtant 150 Miles. The 16th before, we had ſent our Agent, with the Agent of the *Batchelor*, aboard the *Duke*, to demand all the Gold, Plate, Pearl, and Jewels; and, in Caſe of Refuſal, to proteſt againſt the Officers for refuſing to deliver the ſame; but theſe being more private Differences, we ſhall paſs them by for the future, as has been done before. The ſame Day we receiv'd four Hogſheads of Beer from on Board the *Commodore*. We had above 40 ſick at this Time in our Ship; and the 18th in the Morning Mr. *Duck*, another of our Mates, dy'd.

From the 19th the Wind continu'd variable, from the *E. N. E.* to the *S. W. July* 21. a Frigat was diſpatch'd with Letters of Advice for *Amſterdam*. We kept ſounding when upon the Banks, and found all agreed well with our Charts. *July* 23. at Eight in the Morning ſaw the Land bearing about *S. E.* by *S.* diſtant four Leagues, and ſoon after ſeveral Boats with Pilots. We made a Waift, and one of them preſently came Aboard us. From *Shetland*, to the *Texel*, I reckon the Courſe is near *S. E.* by *S.* Diſtance 160 Leagues. At Seven in the Evening, God be prais'd, we came to an Anchor in the *Texel*-Road, having before ſaluted the *Commodore*, who lay off to ſee the Ships go in for *Helvort Sluice*, and other Places. We were on our Paſſage from the Cape of *Good Hope*, to the *Texel*, 3 Months and 17 Days.

Arrive in the Texel.

July 24. 1711. in the Morning, the Flag and *India* Men bound for *Amſterdam*, weigh'd.

We

We saluted the Admiral with nine Guns, and then went ashore, to get Refreshments for our Men. The 27th receiv'd a general Letter from our Owners, which may not be unacceptable to such as desire to see all that relates to our Voyage, and is therefore here inserted.

Bristol, June 6. 1711.

SIRS,

WE *have receiv'd several of yours from sundry Places, particularly that of the 7th of* February *last, from the Cape of* Good Hope, *by the* Oley *Frigat, which is arriv'd in* Ireland. *One of the Super-cargo Men came Post for* London, *and we had the Letters by Express, on* Sunday *the 27th of* May *last, with the agreeable News of the* Duke, Dutchess, *and* Batchelor's *safe Arrival there.*

By the Council's general Letter, your Resolution seems for Holland, *unless you hear of a Peace, or meet with an* English *Convoy. The War continues, and Convoy is doubtful. Upon Receipt of yours, we have consulted, and writ to sundry Friends in* London, *what proper Methods must be taken, should you arrive in* England. *We have also writ to several Friends in* Holland, *to be fully inform'd how to proceed, should you arrive there. We cannot yet expect Letters from* Holland, *but have receiv'd sundry Advices from* London; *all which confirm the* East-India *Company are incens'd against us, and have appointed a select Committee to inspect their Charter, as to their Privileges, Bounds,* &c. *and are resolv'd to take all the Advantages they possibly can against us. We doubt not but that you have acted with all due Precaution Aboard, but there may be Danger of offending at* Home. *Therefore we have writ divers of these*

Copies

Copies to several Ports, viz. Amsterdam, Rotterdam, *and five or six Ports in* Ireland, *that it is our Opinion and Order, that at your Arrival in any secure Port in* Holland *or* Ireland, *you dispatch away Advice, as soon as possible, by Express or otherwise, and remain in Port 'till farther Orders; and particularly to take Care that nothing be landed out of your Ship or Ships, that it may not be in the Power of any Informer to lay the least Accusation against you; for we lye liable both to the* English *and the* Dutch *East-India Company upon any Mismanagement, and they are resolv'd to give us all possible Disturbance. And since it has pleas'd God to bless you and us with probable Success, after your long and dangerous Voyage; and since all your Company, from the Captain to the lowest Mariner, will have a good Interest therein, we doubt not but that you and they will be all unanimous to preserve their own and our Interest, and not commit any rash Act to expose and hazard the whole; it being our Resolution, that every Person Aboard shall have a just and faithful Distribution of his Shares and Wages, and all Encouragement that can be expected. We do recommend it to you, that you read this Letter and Order to your respective Crews, that every one may be appriz'd, that if either Officer, Sailor, or any who shall come Aboard of you, do carry any Goods ashore, for Sale or otherwise, the whole is forfeited and lost; and no doubt, but there will be some employ'd by the* East-India *Company to insnare some of you, who will use all imaginable Art for that Purpose. God send you safe to your discharging Port, that we may have a joyful Meeting, is the Desire and Prayers of*

Your loving Friends,

John

John Batchelor,	Charles Shuter,
John Hawkins,	James Hollidge,
Thomas Clement,	John Romny,
Thomas Goldney,	Laurence Hollifter.

Other Letters are omitted, as not material in the Main to the Publick, they generally contain'd frefh Precautions for fecuring our felves, upon Miftruft of the Defigns of the *Eaft-India* Company. Therefore a ftrict Watch was kept Aboard every Ship, to prevent the carrying of any Thing afhore; and no Perfon whatfoever permitted to come Aboard, for Fear of a Seizure.

The *Dutch* Men fhipp'd at *Batavia*, being uneafy, I had fent up their Accompts to *Amfterdam*, in order to their being difcharg'd; but on the firft of *Auguft*, 18 of them ran away in a Boat of the Country. Having receiv'd Letters directing to carry up the Ships into the *Flitter-Road*, where the *Dutch India-*Men ufually ride, we anchor'd there on the 4th. The fame Day the *Englifh Eaft-India* Men fail'd with Convoy for *England*. The 5th Mr. *Hollidge*, and others of the Owners from *England*, came Aboard, in their Way to *Amfterdam*, each Ship faluting them with 11 Guns at their coming and going. The 11th, moft of the principal Officers in the three Ships went afhore with Mr. *Hollidge*, and the other Owners, to the chief Magiftrate on the *Texel* Ifland, where moft of the faid Officers made Oath to a fhort Journal or Abftract of our Voyage round the World, and that, to the beft of their Knowledge, they had not traded in *India* for any Thing but Neceffaries and Provifions. The 12th, we all return'd Aboard our Ships, and held a

Council

Council Aboard the *Duke*, where, in Confideration of the long Time our Officers and Men had been from *Europe*, it was refolv'd to fupply them with fome Money to recruit themfelves afhore, 20 Gilders to every Sailor, 10 to a Land-man, and to every Officer proportionably to his Neceffities; for which, and for furnifhing Provifions, &c. for the Ships, Mr. *Hollidge* was defir'd to receive the Value of 1800 Pounds Sterling in Gold and other Treafure, to be exchang'd at *Amfterdam*; however, upon fecond Thoughts he did not, but took up Money for the Ufe of the Ships, and gave Bills for the fame.

The 13th, receiv'd another general Letter from our Owners in *England*, promifing to do Juftice to us all, recommending Unity, and directing us to continue there, as not knowing where we fhould unlade, underftanding that the *Englifh Eaft-India* Company defign'd to feize our Ships.

Auguft 14. 1711. moft of our Men fign'd the following fhort Account of our Voyage, being very near the fame that feveral of the Officers had fworn to at the *Texel*.

We whofe Names are under-written, being Sea-men belonging to the Ship Dutchefs *of* Briftol; *do hereby teftify and declare, and are ready, if Occafion require, to make Oath, that what is contain'd in this Paper, is in every Particular true, to the beft of our Knowledge.*

AUguft 1. 1708. we fail'd from *Briftol*, and arriv'd at *Cork* the 5th of the fame Month, where we took in Provifions and Men for our Voyage.

September 1. 1708. fail'd from *Cork*, and arriv'd near *Teneriff* the 18th of the fame Month; on which Day we took a fmall *Spanifh* Bark, call'd the *St. Philip* and *St. James*, *Antony Hernandes* Mafter, come from *St. Cruz*, and bound to *Fuerte Ventura*; which Bark we ranfom'd for fome few Neceffaries and Provifions, fhe being of little Value. The 21ft of the fame Month, we departed from thofe Iflands, and arriv'd at the Ifland of *St. Vincent* the 30th of the fame Month, where we wooded and water'd Sail'd from thence the 7th of *October*, and arriv'd at the Ifland *Grande*, on the Coaft of *Brazil*, the 21ft of *November*, where we again wooded, water'd, and clean'd our Ships. Sail'd from thence the 1ft of *September* 1708, bound for the *Pacifick* or *South Sea*, and arriv'd at the Ifland *Juan Fernandes* in the faid Sea, the 1ft of *February* 1708, where we wooded, water'd, and clean'd our Ships again. Sail'd from thence the 13th of the fame Month, bound to the Ifland *Lobos de la Mar*. In our Paffage thither, we took near the faid Ifland, on the 16th of *March* 1708-9, a fmall Bark, call'd the *Affumption*, *Anthony Villegas* Mafter, bound from *Guayaquil* to *Santa*, Burthen 16 Tuns, which we ranfom'd, and arriv'd at the faid Place the 17th of the faid Month. The 26th of the faid Month, in the Latitude of 7 Deg. 12 Min. *South*, we took a fmall Ship call'd the *St. Jofeph*, about 50 Tuns, laden with Planks from *Guayaquil* to *Lima*, which we ranfom'd at *Guayaquil*.

On the 2d Day of *April* 1709, in the Latitude of 6 Deg. 16 Min. *South*, we took a Ship call'd the *Afcenfion*, *Jofeph Morel* Mafter, bound from *Panama* to *Lima*, Burthen about 450 Tuns, laden for the moft Part with Timber,

fome

some Bale-Goods, and 72 Negroes; the Ship was ransom'd at *Gorgona*. On the 3d of *April*, in the Latitude of 6 Deg. 14 Min. we took the Ship *Jesus Maria Joseph*, *John Gustelius* Master, Burthen 35 Tuns, bound from *Guayaquil* to *Chancay*, which was given to the Prisoners to carry them to *Guayaquil* ashore. The 15th of *April*, in the Latitude of 4 Deg. 8 Min we took the Ship *Havredegrace*, *Joseph de Arisabalaga* Master, bound from *Panama* to *Lima*, Burthen 250 Tuns, laden with Bale-Goods and 74 Negroes, which Ship we fitted with 20 Guns, intending to bring her to *Europe*. The 16th, in the Latitude of 3 Deg. 20 Min. *South*, we took a small Shallop, Burthen 15 Tuns, laden with Provisions, bound for *Guayaquil*; which Shallop we sunk, and put the Prisoners ashore.

The 21st of *April* we took a Vessel call'd the St. *Francis*, *Simon Jacob Debreves* Master, bound from *Santa* to *Guayaquil*, Burthen 40 Tuns, laden with Flower. The Prisoners were turn'd ashore at *Guayaquil*, and the Vessel ransom'd at the Town of that Name.

On the 22d we took in the River of *Guayaquil* two large Ships at Anchor, and five small Barks, without Men, with a small Parcel of Bale-Goods on Board one of the Barks; all which were ransom'd at *Guayaquil*. The same Day at Point *Arena* we took a Vessel, which the Men had quitted, Burthen 50 Tuns, which was afterwards taken by the *Spaniards*. On the 23d of *April* we landed with about 180 Men, and took the Town of *Guayaquil*, and continu'd Masters of it 'till the 1st of *May* following; which Town we ransom'd, and sail'd the same Day. On the 5th of *June*, in the Latitude of 2 Deg. 36 Min. *North*, we took a small

small Vessel, call'd the *St. Thomas de Villa Nova* and *St Dimas*, *John Navarro* Master, bound from *Panama* to *Guayaquil*, Burthen about 90 Tuns, laden with Iron, Pitch, Tar, and some dry Goods, ransom'd at *Gorgona*. The 8th of the same Month, Latitude 3 Deg. 0 Min. *North*, we took a small Vessel, call'd the *Golden Sun*, *Andrew Henriques* Master, bound from Port *St. Joseph*, to *Guayaquil*, about 30 Tuns, in Ballast, which we gave the Prisoners to put them ashore; which Day we arriv'd at *Gorgona*, and continu'd there fitting our Ships 'till the 8th of *August*. On the 18th of the same Month, in the Latitude of 1 Deg. 9 Min. *North*, we took a Ship, call'd the *Conception*, *Francis Salmon* Master, bound from *Panama*, to *Lima*, Burthen 60 Tuns, in Ballast, given the Prisoners to carry them ashore; at which Time we sail'd for the Islands *Galapagos*, and stay'd there a few Days, and thence sail'd to the *Tres Marias*, on the Coast of *Mexico*, Latitude 21 Deg. 30 Min. *North*, where we fitted our Ships, and thence proceeded to the *S. E.* End of *California*, where we cruis'd 'till the 22d of *December* 1709. on which Day, in the Latitude of 22 Deg. 40 Min. *North*, we took, in sight of Cape *St. Lucas*, at the *S. E.* End of *California*, the Ship *Encarnacion*, Sir *John Pichberty* Commander, bound from *Manila* to *Acapulco*, Burthen about 400 Tuns, laden with several Commodities. The 10th of *January* 1709-10, we sail'd from Cape *St. Lucas* with the said *Manila* Ship, now by us call'd the *Batchelor* Frigat, and the Ship *Havredegrace*, by us call'd the *Marquis*, with Design to make the best of our Way for *Europe*.

Round the World.

On the 12th of *March* 1709-10, we arriv'd at *Guam*, one of the *Ladrones* Islands, where we took in Refreshments for our Men, such as the Island afforded. Departed from thence the 21st of the said Month, and arriv'd at the Island *Bouton* the 28th of *May* 1710, where we wooded and water'd; and sail'd from thence the 8th of *June*, and arriv'd at *Batavia* the 20th of the said Month, where we made Application to the Governor for Liberty to purchase Provisions for our Men, and Necessaries for fitting our Ships; which, seeing our Commissions and Journal of our Voyage, they granted. We continu'd there fitting our Ships, which were very much disabled and Worm-eaten, 'till the 14th of *October* following; and examining the *Marquis*, found her Bottom so much eaten by the Worms, that it was impossible to bring her to *Europe*; on which Account, we took out what Goods were on Board of her, and put them on Board the Ships *Duke*, *Dutchess*, and *Batchelor*, and most of the Guns and Materials, and then dispos'd of the Hull for 575 Dollars, which Money was expended for Provisions for our Men, and Necessaries for our Ships. From *Batavia* we sail'd the 14th of *October* for *Java-Head*, where we wooded and water'd. Sail'd thence the 25th of the same Month, and arriv'd at the Cape of *Good Hope* the 29th of *December* 1710, and continu'd there for the Company of the *Dutch East-India* Fleet 'till the 6th of *April* 1711, and arriv'd at the *Texel* the 23d of *July* following.

We also farther make Oath, that we went out as private Men of War, and not as Trading-Ships, and that no Sort of Merchandize was shipp'd on Board the said Ships to

trade

trade withal; and that we have not been in any Place or Places, or Islands in the *East-Indies*, more than what has been above express'd; and that we drove no Trade, nor made any Purchase at *Bouton* or *Batavia*, or in any Part of the *East-Indies*, more than for Necessaries and Provisions.

Sign'd by most of the Men in the Ship.

After this, continu'd here, without any Thing remarkable, 'till the 30th of *September* 1711, when we sail'd from the *Texel*, under Convoy of her Majesty's Ships, *Essex*, *Canterbury*, *Medway*, and *Dullidge*, sent over for that Purpose from *England*. Came to an *Anchor* in the *Downs*, *October* 2. 1711, at 10 in the Morning. *October* 13. the *Duke* and *Dutchess* came up to *Erif*, the *Batchelor* having been there some Time before. At this Place all the Ships moor'd, and continu'd 'till unloaded.

The End of the Voyage round the World.

A DESCRIPTION

OF THE

Sea-Coasts, Head-Lands, Soundings, Sands, Shoals, Rocks, and Dangers; the Bays, Roads, Harbours, Rivers, Creeks, Ports, and Sea-Marks in the *South Sea*, from the Port of *Acapulco*, to the Streights of *Magellan*; shewing the Courses and Distances from one Place to another, the setting of the Tides and Currents, and the Winds generally reigning; with exact Draughts of the Coast, the Bearings, and of several Ports.

The whole translated and copy'd from the *Spanish* Manuscript Coasting-Pilots, gather'd from the Experience and Practice of that Nation, for 200 Years on those Seas.

CHAP. I.

The Sea-Coasts, &c. from the City of Panama, *on the* Isthmus *of* America, *to* Callao, *which is the Port to the City of* Lima, *Capital of* Peru.

THE City of *Panama* is in 9 Degrees of *North* Latitude, a Place of great Trade, where Ships lade for *Peru* and *New Spain*.

Two Leagues *S. W.* from *Panama*, is Port *Perico*, form'd by three Islands, which are a bare League from *Rio Grande*, or the Great River, *N. by W.* and *S. by E.* In the Mid-way from *Panama*, to Port *Perico*, is a dangerous Rocky Shoal, where some Ships have perish'd; and others have struck, and lost their Rudders. From *Perico*, to the said Shoal, is a League; and they bear from one another *N. E.* and *S. W.* and the Shoal with the Hill of *Paitilla North* and *South*. Seven Leagues *E. S. E.* from *Panama*, is the Island of *Chepillo*, half a League in Compass, all wooded with Fruit-Trees, and abundance of Plantans. It lies near the Continent, and on the *South* side of it, is deep Water; but the *North* is so shoal, that small Boats cannot pass, and on it there is fresh Water.

Between the Island of *Chepillo*, and the City of *Panama*, are three Rivers which fall into the Sea. The first the River of *Juan Diaz*, two Leagues from *Panama*; the next that of *Tuguman*, two Leagues farther; and that of *Pancora*, two Leagues beyond the last. The Tide goes up them, they are shoal next the Sea, and shaded with Mangroves.

Half a League to the *Northward* of the Island *Chepillo*, is the great River *Bayana*, which runs up to *Chepo*, where there are many Farms, and much Timber.
From

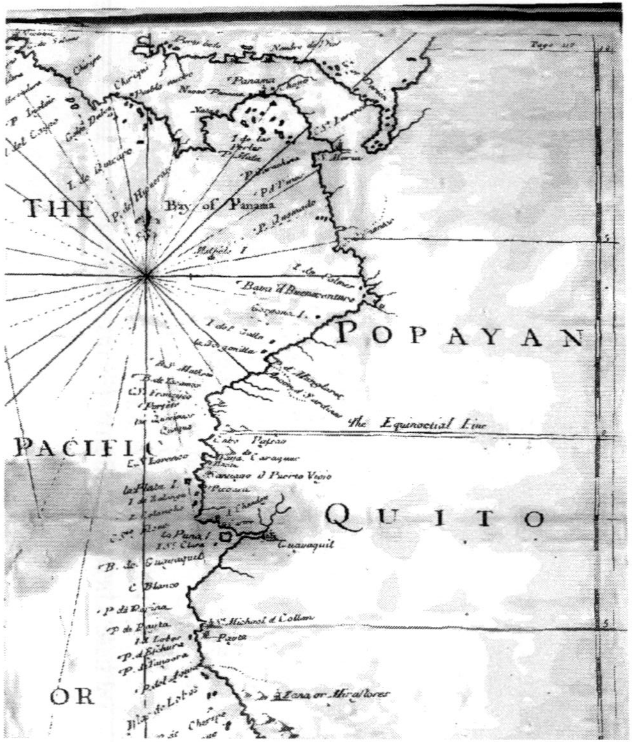

THE
Bay of Panama
Panama
PACIFIC
OR
POPAYAN
QUITO
the Equinoctial Line
Guaiaquil
C Blanco
Bata d'Buenaventura

pillo, is the great River *Bayana*, which runs up to *Chepo*, where there are many Farms, and much Timber.

From

From the Island *Chepillo*, to the River *Maestra*, and Point *Manglares*, or Mangroves, five Leagues, and in that Way three Rivers, being those of *Chinilla*, *Lagartos*, and that of *Pariga*. At the River of *Maestra*, and Point *Manglares*, or Mangroves, there are some dangerous Shoals, running out to Seaward, which are to be carefully avoided; and those who are going in for the Shore, must found all the Way, and when they come into six Fathom Water, turn off short for the Island *Pacheca*, lying *East* and *West* with Point *Manglares*, or Mangroves; for close to that Island there is deep Water. All the Coast between the River *Maestra*, and the Island *Chepillo*, is flat, and full of Shoals, which run farthest out to Sea at the said River and Point.

The Island *Pacheca* is 11 Leagues from *Panama*, N. W. and S. E. and 11 Leagues from those of *Perico* E. S. E. and W. N. W. This Island *Pacheca*, and that of *Chuche*, lie N. N. E. and S. S. W. four Leagues distant. The Island of *Chuche* is the biggest of all the Pearl Islands, next to the *King's* great Island, and the Port where the Boats anchor at it, is on the *North* side. There are abundance of Mice on it; but the Water is deep: 14 or 15 Fathom within a Bow-shot, it has fresh Water, and on the *South* side there are seven Rocks above Water, or small Islands, close by one another. *Chuche* is 15 Leagues from *Panama*, North and *South*, 12 from the Island *Otoque*, N. W. and S. E. and 12 from that of *Taboga* N. N. W. and S. S. E.

Taboga and *Otoque* bear N. N. E. and S. S. W. four Leagues distant. *Taboga* is above a League in Compass, high and woody, the Port on the E. S. E. side, and in it is a Creek of fresh Water running down into the Sea, where the Boats take it up. This Port is deep, and has good Hold for Anchors, and about it are several other Creeks of fresh Water. Near this Island is another smaller, call'd *Taboguilla*, or little *Taboga*. Ships may pass between them, keeping close to

Taboga;

Taboga; for *Taboguilla* has a rocky Shoal, where the Sea is often seen to break. Neither must you come too near the Head-land of *Taboga*, when getting out of the narrow, because, if the Flood rises, it will drive you upon the Head-land.

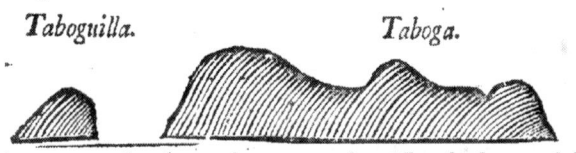

Taboguilla.   *Taboga.*

As you pass along between the Islands beyond *Pacheca*, facing *West*, you will see *Taboga*, which shews thus; as also *Taboguilla*, much less than the other, appears as above.

At *Isla del Rey*, or the *King's Island*, on the S. W. side, there is deep Water, but shoal on the N. E. as far as the River *Mahe*, at the Mouth whereof is a small Island; from the N. W. side whereof several Shoals run out; be sure to keep clear of them. Two small Leagues short of the River *Mahe*, is that of *Chiman*, with a little bare Island or Rock at the Mouth of it, which you run up close to, because there is deep Water, but cannot pass to the Landward of it, by Reason there are many Shoals. Between this River of *Chiman*, and that of *Maestra*, is *Rio Hondo*, or deep River, by another Name call'd *Boca tuerta*, or wry Mouth, very deep, and so full of Windings within, that the Sea cannot be seen, having eight Fathom Water. At the Mouth of it, there are some Sand Banks; and there must be Care and Skill to get into it. All the Way from the River *Chiman*, to that of *Pena Oradada*, or the pierc'd Rock, which is two Leagues, there are good Trees, call'd *Marias*, for Masts and Yards, as also Cedars, Oaks, and Medlar-Trees.

From the small Island of *Chiman*, to Cape St. *Lawrence*, there is deep Water, and anchoring in 10 and 12 Fathom, but come no nearer to the Land, because there

there are many Rivers and Shoals. At Cape St. *Lawrence* you may anchor, for it is clean Ground. All the Coast above-mention'd, is low, and cover'd with Mangroves, as far away as the Bay of St. *Michael*, in which there are some small Islands, and considerable Rivers, where there is safe anchoring, without Danger of Winds blowing in, in above 15 Fathom Water.

The Island *de la Galera*, or, of the Galley, and Point *Garachine*, bear *E. S. E.* and *W. N. W.* and two small Leagues short of the Island, and somewhat above three from the Point, is the Shoal of St. *Joseph*. From the Island *Galera*, to Point *Garachine*, is five Leagues. On St. *Joseph*'s Shoal there is two Fathom Water in some Places, and three in others, and it bears with Point *Garachine East* and *West*, and with the Island *Galera* almost *E. by N.* and *W. by S.* Ships may pass between the Island and the Shoal, keeping closest to *la Galera*, as also between the Continent and the said Shoal. At the Ebb the Water sets strong upon the Islands. Two Leagues to the Offing *N. E.* and *S. W.* with the Island *Galera*, is a Bank the Sea uses to break on, but there is much Water about it; and on the *S. W.* side of the Island *Galera* close to it, there are rocky Shoals. The Sea runs violently between these Islands at the Ebb and Flood, from *August* to *October*, which are the three Months when the *S. W.* Winds blow hardest; and then you may sail without the Islands, because there is Room for long Trips upon tacking. The rest of the Year you may go within them, because the Season is calmer, and there are Land Breezes, and you may come to an Anchor. Point *Garachine* is small from the Bottom to the Top, somewhat full of Hillocks, and a Ridge on the Top, and in it a small Break, call'd *Sapo*.

Vol II.   I   Point

A Description of

Point Garachine.

When this Point *Garachine* at (o) bears *E.* by *N.* it shews thus. The next Bearing joining the two Lands, where the Crosses are, is the Coast as it runs along to Port *Pinas*; all which may be seen in clear Weather.

The Land of Port *Caracoles.*

Joining this with the other above, as is there said, it represents the Land as it rises, from Port *Caracoles*, to Port *Pinas*, which may be seen in clear Weather, the Cross being eight Leagues to the *Eastward*.

The South Sea Coasts.   115

Point Garachine.   Sapo.

As you come from the Offing, to make the Land of *Garachine*, the Hill call'd *Sapo* above-mention'd bearing *N.* by *E.* it shews thus. If the Weather be clear, you will see Port *Pinas* towards the *S. E.*

Point Garachine.   Sapo.

When the Hill *Sapo* bears *E.* by *N.* it shews thus, seven Leagues out at Sea.

From Point *Garachine*, to Port *Pinas*, is seven Leagues *N. W.* and *S. E.* and by the Way, in a small Bay form'd by the Coast, is Port *Caracoles*, above-mention'd. Port *Pinas* is the highest Land, and most copling of any on that Coast; and on the Top of the highest Mountain are certain Brakes, which appear plain when Port *Pinas* bears *N. E.* A little without the Mouth of this Port, you'll see two little Isles, or Rocks, close to one another, on either side of which, Ships may pass in or out; but the best Way is between them. To the *S. E.* of this Port, there

I 2

there are four or five little Isles, which you are not to come too near. When in this Port, you'll see a large Bay, every where deep Water, and safe anchoring in any Part; and at the *N. E.* End of the Bay, is a sandy Shore, on which you may lay the Head of your Boat, going up On the right Hand, is the Pine Grove, and on the left a River of Salt Water; but half a League up it, you'll find fresh Water, which runs down from the Top of the Mountain. Here are Masts and Yards, and Places to careen; and farther up, another little Bay is form'd by the Sea. At this Port *Pinas* there are sometimes *Indians* in Arms, for which Reason it is not safe to go ashore unprovided; and when up the River, take Heed your Boat be not left dry.

From Port *Pinas*, to *Morro Quemado*, or burnt Headland, 12 Leagues *N. W.* and *S. E.* the Land lowering all the Way from the high Mountains of Port *Pinas* to the *S. E.* 'till you come to *Morro Quemado*; and about the Mid-way, there are some flat Brakes among the Mountains, which may be seen when they bear *N. E.* Three Leagues short of *Morro Quemado*, is a small Bay, and at the lowest Land is a River, and abundance of Coco-Nuts. *Morro Quemado* is high Land; and when bearing *East*, at a considerable Distance, appears even on the Top; but when near, there is a Brake, which makes a large round Head, upright next the Sea, and without that Head is a high Rock.

Morro Quemado,

Bearing *E.* and *E. S. E.* shews thus.

Morro

Morro Quemado.

Morro Quemado bearing *N. N. E.* shews thus, and the Land rises from the Cross towards the *S. E.* as far as *Puerto Quemado*.

From *Morro Quemado*, to *Puerto Quemado*, or burnt Harbour, which is a high Head-land, and to the *S. E.* of it are three or four small high Rocks, or bare Isles; the Distance between *Morro Quemado*, and *Puerto Quemado*, three Leagues.

Puerto Quemado.

When the Height of *Puerto Quemado* bears *E.* by *N.* it shews thus.

At this Head-land of *Puerto Quemado* begin the *Anegadas*, or overflown Islands, and run along 'till within six Leagues of Cape *Corrientes*, or Currents. At these *Anegadas* is a Bay of low Land, full of Hillocks, which, when out at Sea, look like so many Islands parted from each other, and much Depth of Water there is there. *Note*, That there is no anchoring all along this Coast, from Port *Pinas*, to Cape *Corrientes*; nor is it safe going in too close with the Shore, lest the Wind should start up athwart, and, together with the Sea, run the Ship a-ground, especially when the *Vendabales*, that is, the *South West* Winds prevail; but if the Trade-Winds reign, you may venture near the Coast, for then the Wind and Currents bear off. Observe, that all along this Coast, the *Indians* are not subdu'd. At the End of the *Anegadas*, is a large Bay, well land-lock'd, and good anchoring; and from this

Bay the Land runs somewhat towards the *S.W.* and the Point form'd by the Land coming from Cape *Corrientes*, and from the Bay on the other Side, is call'd *las Salinas*, or the Salt-Pits, which coming from the Seaward, looks like the Point of *Garachine*, bating that the high Land is not so lofty as that of the other call'd *el Sapo*.

From *Morro Quemado* to Port *Pinas*.

The Head-land, or Point at the Cross, is that which advances from *Morro Quemado* towards Port *Pinas*; and when the little Head-land at (o) bears *East*, it shews thus.

The Great *Anegada*.

The Great *Anegada*, bearing *E. N. E.* when you are eight Leagues out at Sea from it, shews thus.

The Great *Anegada*, bearing *E. N. E.* at a greater Diſtance, ſhews thus.

Point *Salinas*, in 5 Deg. 45 Min. *North* Latitude.

When the Point at the Croſs bears *Eaſt*, the Land of *Salinas* ſhews thus.

From this Point *Salinas*, or, of Salt-Pits, to Cape *Corrientes*, or Currents, ſix Leagues; and from Port *Quemado*, to Cape *Corrientes*, 29 Leagues. This Cape *Corrientes* is a high round Head-land, on the Coaſt, with two Hillocks on it, cloſe to each other. The high Head is half a League up the Land, and when it bears *S. E.* looks like a high round Iſland. From Cape *Corrientes*, to the River *Nionimos*, 10 Leagues *N. W.* and *S. E.* the Coaſt low, the River large, and has two Mouths. The Banks of it inhabited by *Indians*, who are ſometimes at Peace, and other whiles at War; but there is no truſting them, for they go about in Canoes, robbing the Boats they meet. Oppoſite to this River, is the Iſland *Palmas*, low, and about it are many Shoals, and moſt on the *S. W.* ſide. From the River of *Nionimos*, to the Bay of *Buena Ventura*, or, good Fortune, 10 Leagues; to ſail for which River, you are to make into a large Bay, call'd alſo *da la Buena Ventura*, or, of good Fortune, into which three Rivers and many Creeks fall. The Rivers are, thoſe of *Lagua*, of *Buena Ventura*, and of *Oſtiones*. The Bay has little Water, and many Shoals. From the River *Oſtiones*, a Shoal runs half Way into the Bay.

A Description of

Cape *Corrientes*, in 5 Deg. 15 Min. *North* Latitude.

Thus it shews, when bearing *South*.

Cape *Corrientes*.

Thus it shews, when coming from the Seaward, and bearing *E.* and *N. E.*

Cuchilla de Sotavento del Cabo de Corrientes, or the Leeward Ridge of Cape Currents.

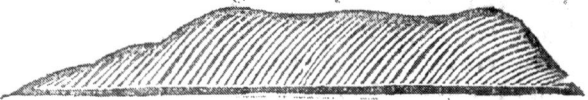

Coming for the *Anegadas*, or Port *Quemado*, and facing to the *S. E.* in fair Weather, this Ridge may be seen within Cape *Corrientes*, at 10 Leagues Distance from the Coast.

In the Bay of *Buena Ventura* you may anchor under a Head of Mangroves, on the right Hand, which is an Island made by the River of *Ostiones*, and call'd *Realejo*; you are there to anchor before a small Creek, which, when moor'd, will be a-head of you. The Shoal I mention'd above to be in the Bay, does not appear but at low Water. If you would go up to the Fort, find out the Channel; and coming into the Bay, as far as the Land which runs to the *N. W.* you will find from four to five Fathom Water, sounding all the Way, and come not too near the Land, for there are many Shoals. This Land trends to the *N. W.* and has no Mangroves; but the Continent

makes

makes an indifferent upright Head, on which are some
white and red Spots ; and farther to the *N. W.* is the
Island *Palmas*, making a sharp Point.   Near the Head
last mention'd, is a little round Island, with some
Trees on it, not to be seen out at Sea, because it is
close to the Shore ; but when you are in the Bay, and
when a-stern of you.   It bears with the Mouth of the
River *Buena Ventura, E.* and *W.* somewhat inclining
to *N. W.* and *S. E.* and in the same Manner with the
Island call'd *Realejo*.   The River of *Buena Ventura* has
but little Water ; so that no Ships can go up it, but
only the Trading-Boats.   When you come into the
River, you will see a Tree in the Sea, which you are
to leave on the Right Hand, to sail up ; and then
you'll see a Creek on the Left, go not up it, for there
is no Way out ; and when you come to the four
Mouths, go not up the first, on the Right Hand, because there is no passing, and you'll be a-ground ; but
make up the second, which is safe, and tho' it seems
narrow, is not so.   Going up, you'll see a Creek on
the Right, call'd *de Pero Lopez*, run not into it, because it leads into the Sea.   Above it is another on
the Right, call'd *de los Piles*.   Keep always to the
Left, 'till you come to *Puerto Viejo*, where there are
many *Guayavas* and Lemons, from the Time when this
Place was inhabited. From *Puerto Viejo*, is two Leagues
to the Fort, opposite to which you are to anchor.
The Marks to know this Bay of *Buena Ventura* out at
Sea, are, low Land next the Sea, and high copling up
the Country.   If the Weather be fair, you'll see on the
Inland a high Ridge, and on it a Sort of Peek, and
to the *Southward* of this Ridge, another thicker Hill,
call'd *de las Minas*, or, of the Mines, which seems to
be separated from the high Lands ; and still to the
Southward of it, is another round Hill, like a Sugar-
Loaf.   If it be clear above, you'll see all this Land
plain.   Many considerable Rivers fall into this Bay,
among which are those of *Piles* and St. *John*. There
are

122　　　*A Description of*

are many Shoals running two Leagues out into the Sea; so that it is a dangerous Coast.

Buena Ventura Bay, in 3 Deg. 45 Min. *North* Latitude.

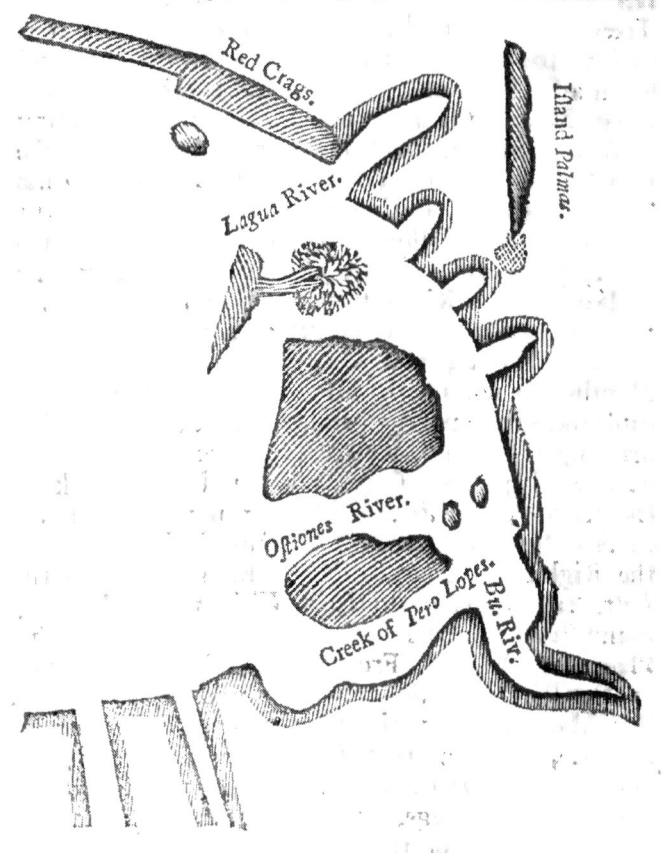

If you will pass between the Island *Palmas* and the Continent, it is deep, and you may turn it in, for there is 12 Leagues Distance.

From the Bay of *Buena Ventura*, to the Island *Gorgona*, 20 Leagues *N. E.* and *S. W.* The Island is high Land, about two Leagues in Compass, lying in Length
　　　　　　　　　　　　　　　　　　　　　　　N.

N. E. and *S. W.* and has fresh Water next the Continent at a sandy Shore, where there is much Water, about 50 Fathom, a dangerous anchoring Place when the Wind is *S. E.* for it blows full in. Here are Plantans, as also on the small Island. When you design to anchor on the *S. E.* side of this Island *Gorgona*, keep clear of the small Island, and a Parcel of Rocks you'll see there. From the Island to the Continent, which is there low and shoal, is four Leagues. This Island from the Offing, when it bears from *S. E.* to *N. E.* shews two Brakes; next the *N. E.* Point it is lower, to the *S. W.* it makes a thick Head-land, and in the Middle forms a round Peek rising higher than all the rest of the Island, which, seen at a Distance, looks like a high Rock, or small Hillock in the Sea; and when you draw nearer, shews as below.

Gorgona Island, in 3 Deg. 15 Min. *North* Latitude.

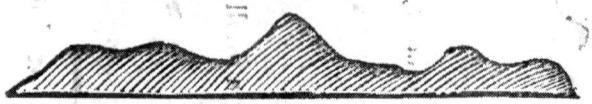

When it bears from *S.* to *S. E.* you'll see the little Island parted from the great one, and the low Point, which runs from *Gorgona* somewhat to the *S. W.* If the Weather be clear, when you look towards the *S. E.* you'll see high Mountains, which usually look white.

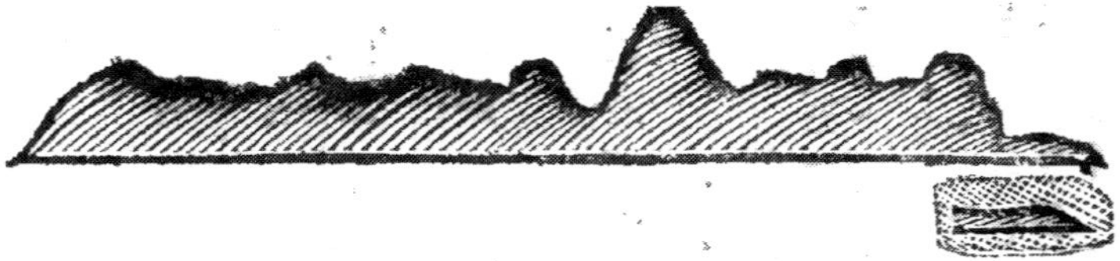

124 *A* Description *of*

The Island *Gorgona*, at fix Leagues Distance, and bearing from *S. E.* to *N. E.* shews as above.

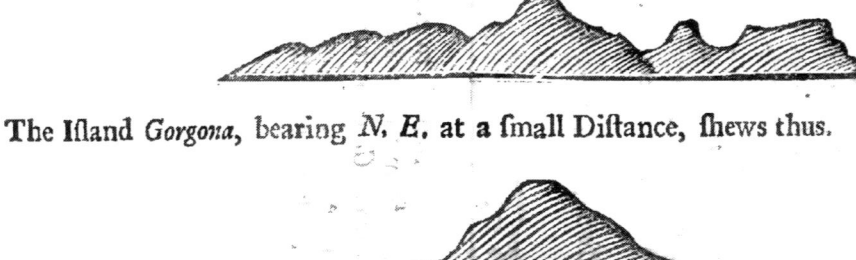

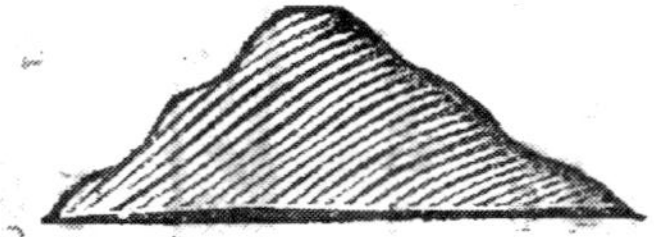

The Island *Gorgona*, bearing *N. E.* at a small Distance, shews thus.

The Island *Gorgona* from the Seaward, at first Appearance on every Side, shews thus. Standing

Standing in from the Offing to make Land, to the Windward of Gorgona, whether you see the Island at a great Distance, or see it not, the Land of the Continent bearing E.S.E. will shew as above.

From the Island *Gorgona*, to that call'd *del Gallo*, or, of the Cock, 24 Leagues, all the Coast low, full of Mangroves, subject to be overflow'd, and full of Shoals, occasion'd by the many Rivers coming from the Continent. The first of them call'd *de Cedros*, or, of Cedars; the second *de las Barbacoas*, near a low Point call'd also of *Barbacoas*, where those *Indians* formerly dwelt, who are now remov'd near the Island *del Gallo*. From this Point *Barbacoas*, to the Island *del Gallo*, is still low Land, like that before-mention'd, cover'd with Mangrove Trees, and full of Flats, and all along it the Shoals run out two Leagues into the Sea. If you happen to ply upon a Wind on this Coast, come not within three Leagues of the Shore; and tho' the Wind be large, hale out, and come not into less than 15 Fathom Water, or you'll be a-ground, when you least think of it. Coming from the Offing, to make this Point *Barbacoas*, you'll see these Mountains up the Inland, and the Sea Coast, when near, very low, and subject to be flooded.

Thus

Thus shews the Point *Barbacoas*, when posited as above.

Short of the Island *del Gallo*, is the River of St. *John*, by others call'd of *Telembi*; and on the *South* side of this River, is a little Mountain, of no great Height, call'd *Morro de las Barbacoas*, or, the Head-land of the *Barbacoas*, because the Inhabitants of the Point above-mention'd planted Colonies here.

Head-land of *Barbacoas*, in 2 Deg. 20 Min. *North* Latitude, bearing S. E. by E.

S. W.

The Head-land of *Barbacoas* to the *S. E.* by *E.* shews thus. You'll see the Island *del Gallo* to the *S. W.* making a Brake, which looks like a great Island and a small one, 'till you draw near, when it appears to be all one, and on the Head-land is a reddish Slough. When nearer, it makes not so many little Brakes, only the *S. W.* Head-land appears cleft on the Top next the Coast.

The South Sea Coasts.

The same bearing East.

N.E.　　　　　　　　　　　　　　　　　　S.W.

When you come from the Offing, and this Head-land of *Barbacoas* bearing *East*, it shews thus; and the Island *del Gallo* is lower on the *South* side. If you view it more to the *N.E.* the *S.W.* Point will appear lower, and that to the *N.E.* higher.

The same bearing E. N. E.

If it bears *E. N. E.* or more towards the *N.E.* distant about eight Leagues, it shews thus, and the Island *del Gallo* seems to be a Piece of the same Continent with the Head-land, and you always see the red Crag or Slough on this Head-land; and there are other small red Crags on the Island *del Gallo*.

The Island *del Gallo* is not so high as the Head-land of *Barbacoas*, and at a Distance has a Brake on the *N. E.* side, which looks like two Islands, that which is towards the *N. E.* being the smaller. The Island lies *N. E.* and *S. W.* and the Point to Windward is lower than that to Leeward. If you coast along upwards, at about six Leagues Distance, you'll see it parted

parted from the Continent, for they are half a League afunder. The Head-land of *Barba-coas* to the *N.E.* is higher than the Ifland. If you come from the Offing, to make this Coaft, note, that from the Ifland *Gorgona*, to the Windward, there is no high Land on the Coaft, befides the Head-land of *Barbacoas*, and this Ifland *del Gallo*.

The Ifland *del Gallo*, in 2 Deg. 15 Min. *North* Latitude.

This Ifland, bearing *S. E.* fhews thus.

The fame Ifland bearing *N. E.* fhews thus.

When this Ifland bears *Eaft*, there appears another Brake to the Windward, with a little upright Head-land, and the Point goes off fharpening to the Windward, and at the End is another little upright Head-land. If you difcover it coming, from above, there will appear no Marks to be obferv'd, by Reafon it is lower than the Head-land of *Barbacoas*, and they are fo pofited, as to appear all the fame Head-land and Continent. On the River *Tenlebi*, on the Continent, there are Mangroves; if you ftand in need of any Mafts, you may anchor on on the *N. E.* Side of this Island *del Gallo*, tho' there is but five Fathom Water, you may go up with Safety, for it is all clean, before a sandy Shore; and if there be Occasion, you may lay the Ship a-ground, and wood and water, just beyond the Island *del Gallo*.

From the *Island del Gallo*, to Point *Manglares* or *Mangroves*, nine Leagues, *N. N. E.* and *S. S. W.* and in the Mid-way between them, is a small Island call'd *la Gorgonilla*, somewhat high, otherwise call'd *Tumaco*, which, when bearing from the *South* to the *East*, shews round, as below.

La Gorgonilla bearing from *South* to *East*.

Opposite to this Island, is a River, where, in case of wanting Water, Ships may anchor at the Mouth, or within it; for there is a clear Bottom. All the Coast is low, and there is safe Anchoring along it. Point *Manglares* or *Mangroves*, is low, very woody, and from the Seaward, looks as if it were subject to overflow. Come not near, for the Shoals run from it two Leagues into the Sea.

Beyond Point *Manglares*, there runs in a large Bay of low Land, call'd *Ancon de Sardinas*, or the Bay of Pilchards, full of Shoals, as far as the Mouth of the River of *Santiago*, where the Land begins to rise. From Point *Manglares*, to the River of *Santiago*, is 15 Leagues, *N. E.* and *S. W.* There is five Fathom Water in the Channel of this River of *Santiago*, and along it there is much Cedar; the Country inhabited by *Indians*, who are not subdu'd. The River is in 1 Deg. 20 Min. Latitude *North*.

From the River of *Santiago*, to the Bay of St. *Matthew*, five Leagues, *N. N. E.* and *S. S. W.* high Land, with red Crags or Sloughs. If you have Occasion to anchor, on Account of some hard Squals, and the Sea's running high, as often happens hereabouts; you may do it under the Shelter of some Points running out from the Coast, between the River of *Santiago*, and the Bay of St. *Matthew*, in what Depth you shall think fit, provided it be not under six Fathom. If you would go into the Bay of St. *Matthew*, you must keep close to the Windward Crags, that the Ebb may not drive you out, or upon the Shoal that is at the Mouth of it. You must anchor in seven Fathom, and not under; and if you want Water, go up the River in your Boat, where you will find it. There was formerly on this River a Town of *Mulatto's*, who sold *Maiz* and Fowl.

The Bay of St. *Matthew*.

The Bay of St. *Matthew* bearing *South*, at a great Distance, shews thus, and you'll see the low Land, full of Hillocks, running to *S.W.*

When

The South Sea Coasts. 131

When you come from the Offing, to view the Bay of St. *Matthew*, and the high Part of it bears E. N. E. eight Leagues distant, it shews thus; and you'll see a little Mountain appear through the Break at the Cross.

From the Bay of St. *Matthew*, to that of *Tacames*, three Leagues. The Land lowers all the Way from St. *Matthew*'s Bay, to the middle of that of *Tacames*, and there begins to rise towards Point *Galera*. If you would water at *Tacames*, look out 'till you see five red Crags, where it is all clean Ground, and deep Water; and opposite to the said Crags, you'll find a Break, where there is Water; but if there should happen to be none, because it is Summer, go about a Musket-shot up the Break, and you'll find Water in a great Pool, where it never fails. *Note*, That between the Bay of St. *Matthew*, and that of *Tacames*, there is a dangerous Shoal somewhat near the Continent, keep clear of it.

K 2   The

When the Break at the Cross bears East, about eight Leagues distant, it shews thus.

132 *A Description of*

The Bay of Tacames.

To N. E. somewhat Easterly. Point Esmeraldas. Crags. Tacames R. Rock. Rocks S. by W. To N.W. by W. Headland N. Point Galera.

From the Bay of St. *Matthew*, to Point *Galera*, or the Galley Point, six Leagues *N. E. by E.* and *S. W. by W.* Point *Galera* is very slender, and sharpens away from *Tacames*. It looks like a Galley that is sunk, with the Keel up, and the sharp Point the Sea forms, represents the Beak. You may anchor in this Point *Galera*, half a League to the Leeward of it, keeping a little off from the Point, because of the Shoals there are under Water. There is Water up the Wood, in a little Brook, which runs not down to the Sea, unless in the Rainy Winter Season.

From Point *Galera*, to Cape St. *Francis*, two Leagues, high woody Land. The very End of the Cape, from a small Distance in the Offing, shews higher, and at a greater Distance looks like

like a great Island. If you coast along this Cape, take Notice that it makes three upright Head-lands towards the Sea; and when you will think to weather the one, you'll see the next; and near the third you'll discover two Rocks above Water close together, and thence you'll see the Land bend inward, forming a Bay of lower Land towards *Portete*.

Cape St. *Francis*, in 1 Degree, Latitude *North*.

When bearing *N. E.* eight Leagues distant, it shews thus.

When you come from the Seaward, to make Cape St. *Francis*, at first Appearance it looks like a high round Island; when the Head-land at the Cross bears *E. S. E.* eight Leagues distant, it shews thus.

Bearing *East*, it shews thus.

From Cape St. *Francis*, to Cape *Pasado*, 22 Leagues, in a direct Run, lying *N.* by *E.* and *S.* by *W.* Immediately beyond Cape St. *Francis*, to the Windward, begins a large Bay from Cape to Cape; go not into it, for there are many Shoals. Five Leagues from Cape St. *Francis*, is *Portete*, or the little Port, the Coast lying *N. W.* and *S. E.* The anchoring is in five Fathom, near the Head-land; but you must found all the Way, because of some Sand Banks, call'd *del Portete*, or of the little Port. There is Water on the left Hand near the Head-land. The Wind here generally blows hard from Noon 'till Night, and then it grows calm, especially from *May* 'till *December*. The Land at *Portete* is high, and then falls away to the *South-ward* of it, where it is low as far as St. *John de Quaques*; and in this Bay of low Land, there are three Rivers call'd of the *Cojimies*, from which there run Shoals three Leagues out to Sea, and at the End of those Flats there are some little Islands. Come not near any Part of the Coast of this Bay; for if the Wind falls, and the Tide flows, you will be drawn up to the Rivers. When you sail from Cape St. *Francis*, or *Portete*, run not too far along the Shore, withoutstanding for the Offing, tho' the Wind be large, 'till you come to *Barrancas Bermejas*, or the Red Crags, which some call *Vasia Borrachos*, which is high Land, with deep Water along the Coast. These *Barrancas Bermejas*, are red and white Crags, which from the Seaward look like

like Heaps of Salt, and are six Leagues to the Windward of the Rivers *Cojimies*. If there be Occasion, keep close up to the Land, leaving nine Crags to the Leeward, and you will then anchor conveniently a Musket-shot from the Land, in 14 Fathom clear Ground. There, about a Musket-shot up a Break, you'll find some Pools, in which there is Water all the Year, and may use any of them. One of these Pools is directly under the Equinoctial.

From *Vasia Borrachos*, to Cape *Pasado*, is six Leagues *N. E.* and *S. W.* from a Point, which makes a little white Head-land, call'd Point *de la Ballena*, or *Whale Point*. Cape *Pasado* is a double Land, that is, one high Land appearing above another, and full of Shrubs on the Top. Just at the very End of the Cape it forms a little upright Hollow, with Crags, which, when near, looks like a Wall, and to the Leeward of it is a small sort of Port, where you may take Shelter, if there be strong Gusts of Wind, keeping the Cape to the *W. S. W.* where you may ride.

Cape *Pasado*, in eight Minutes *South* Latitude.

When it bears *South*, shews thus.

When bearing N.E. it shews thus.

From Cape *Pasado*, to the Bay of *Caracas*, four Leagues, high Land next the Coast, with some white Crags, lying N.W. and S.E. There is no going into this Bay to the Leeward, for there are Shoals in the Entrance, and there is no plying to Windward, but you must keep to Windward to go up to the Crags of *Choropoto*, which are white, and then along the Coast. To put into the aforesaid Bay, you must keep close to the Crags of *Choropoto*, carrying little Sail, always clinging to the Leeward Shore in four Fathom, and four and a half. Within it is dead Water, where neither the Wind, nor the Sea can do any Harm. *Note*, that if the Wind falls calm as you are going up this Bay, you may tow in, but must tow short, that the Sea and Stream may not drive you on the Shoals, which lye before the Bay. When you are to come out, do it to the Leeward, keeping close under the Land that runs to Cape *Pasado*. In this Bay there is Wood and Water, as also Plenty of Cattel and Shell-Fish.

From the Bay of *Caracas*, to the Port of *Manta*, nine Leagues, N.E. and S.W. high Land next the Sea, with some little Breaks and Crags to the River and Town of *Choropoto*, where the Coast begins to lower, and form a Bay 'till you come to *Manta*. About two Leagues short of which, is a low Point, call'd *Punta de Camas*; come not near it, because there are Shoals running out to Sea. Somewhat short of *Manta*, up the Country, is a little Mountain, with some small

Breaks

Breaks on the Top, and fall more to the *Southward* is the Hill of *Monte Chrifto*, which is high, and has a little Break on it, and the Land falls away towards the S. *W*. If you keep to Windward, to put into this Port of *Manta*, you muft found all the Way, becaufe of the Flat which lyes at the Entrance; and provided you keep the little Hill at the Crofs very open, fo as you may fee it above the Coaft, you'll be fafe from ftriking on the Bank; and as foon as the little Hill at the Crofs comes to bear with the End of the Town, you are paft the Shoal, and will advance in fix Fathom Water, to anchor in feven Fathom; and when the Church bears S. *W*. you are in the Port of *Manta*.

Monte Chrifto.

When the high Land of it bears *S. S. E.* fhews thus.

When bearing *E. S. E.* it fhews thus.

When

When bearing *N. W.* thus.

From the Port of *Manta*, to Cape St. *Lawrence*, four Leagues, *E. N. E.* and *W. S. W.* and from *Manta*, which is low Land, the Coast rises by Degrees to the Cape. Note, that at a Point there is in the Middle, with a little Island, there are Flats running out; and short of the Cape, at a Bay form'd by the Coast, there is a Bank running a League into the Sea. Cape St. *Laurence* is high and perpendicular; close by it are two small Islands, or high Rocks, call'd the *Friars*, the one slender and higher than the other. It is all deep Water there.

All along the Coast, between *Manta* and *Santa Elena*, there are Anchoring-Places of a good Depth, and a great Ship may safely pass between the Island *Plata* and the Continent; for there is from 10 to 12 Fathom Water, and there is Anchoring at the *East* Side of it.

The South Sea Coasts. 139

Cape St. *Laurence* in 1 Deg. Lat. *South*.

When bearing *South*, distant four Leagues, shews thus.
When Cape St. *Laurence* bears as above, and the two little Islands or Rocks appear without it, look out to the *S. W.* and you'll see the Island *Plata*, or of Plate. All about it is clean; so that you may pass any Way. From the Cape, to the Island, is four Leagues. On the *South* side of it there are some high Rocks; at first Sight it appears round and high, nearer at Hand it looks like two Islands, and soon after you may see it is one.

The Island *de la Plata*, in 1 Deg. 15 Min. *South* Latitude.

Bearing *S. S. W.* at a small Distance, shews thus.

Bearing

Bearing *N. E.* thus.

Bearing *S. E.* six Leagues distant, thus.

Bearing *East*, distant six Leagues, thus.

From this Island *Plata*, to Point St. *Helena*, 18 Leagues *North* and *South*. From Cape St. *Laurence* Leeward to Port *Callo*, five Leagues *N. W.* and *S. E.* the Land lowering all the Way to the very Port of *Callo*, which is a little Bay, and Ships anchor in it to the Leeward of a little Island, which is to remain to the *Southward*, in six Fathom Water. *Note*, That near this little Island, is a Shoal; and this is a better Port when the Trade-Winds prevail, than that of *Manta*, where there uses to be a rough Sea.

From the Port of *Callo*, to the Island *Selango*, four Leagues *North* and *South*; and between this Port and Island there are two Ports, about a League distant from each other, and they are known by some white Crags to be seen there. The Port which lies most to Leeward, is the deeper; they are both inhabited. The Island *Selango* and that of *Plata* lie *N. N. W.* and *S. S. E.* six Leagues distant. The Land is somewhat high next the Sea, with high Hills up the Country; and along the Coast there are some little Bays and Strands.

From

From the Ifland *Salango*, to the River of *Colanche*, seven Leagues *N*. by *W*. and *S*. by *E*. indifferent high next the Sea, and higher up the Country, being the Mountains of *Picoafa*, which end at the very River of *Colanche*; and if you obferve them from the Offing, they narrow away to the *Southward* into a Ridge. Two Leagues fhort of *Colanche*, you'll fee two little Iflands, or high Rocks, call'd *los Ahorcados*; and two Leagues to the *Southward* of them, is a little Ifland with a Rock by it, call'd the Ifland of *Colanche*. It is deep Water, and there is anchoring in any Part of this Bay, and Ships may pafs by any Part of the little Ifland, for it is all clean Ground. In the River of *Colanche* there is frefh Water, which is carry'd thence to the Town at Point St. *Helena*, which is two or three Leagues thence *N. E.* and *S. W.*

At this Port of St. *Helena*, if Occafion be, there are Refrefhments to be had; the anchoring is before the Houfes, in four Fathom, all about is flat, fo that you will every where find the fame Depth. Remember, not to anchor from the Town towards Point St. *Helena*, for there are abundance of Mice, and fome Shoals.

From this Port of Point St. *Helena*, to the Point it felf, is about a League and a half, the Coaft almoft level with the Sea *N. W.* and *S. E.* The Point it felf is high, and at firft Sight looks like a roundifh Ifland, bare at the Top; but as you draw near, you'll fee a lower Point run out, fharpening towards the Sea.

Point

142   *A Description of*

Point St. *Helena*, in 2 Deg. 20 Min. Latitude *South*.

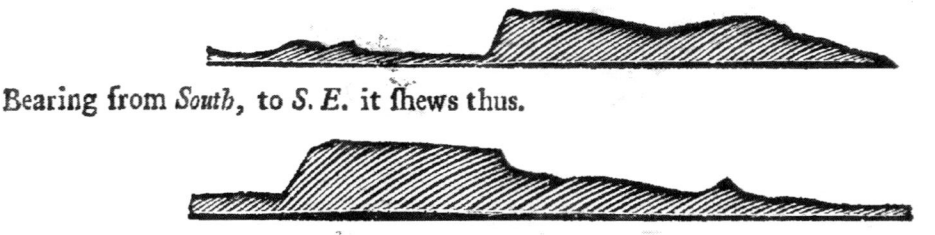

Bearing from *South*, to *S.E.* it shews thus.

Bearing *North*, thus.

To the Windward of Point St. *Helena* runs in a spacious Bay, lying *N*. by *E*. and *S*. by *W*. as far as *Tumbez*, where is the Mouth of the River of *Guayaquil*. From Point St. *Helena*, to *Chandui*, eight Leagues *E. S. E.* and *W. N. W.* low Land, and most of it white Sloughs; and three Leagues to the Windward of the said Point St. *Helena*, is the Point *del Carnero*, with some Rocks like little Islands close to it. Over *Chandui* you'll see some high Hills and Breaks, and still to the Windward, if you are within the Bay, you'll see the Heights they call of the Island of *Puna* and *Chandui*, where is the Mouth of the River of *Guayaquil*. No Ships go up this Mouth, by Reason of the many Shoals.

The

The South Sea *Coasts.* 143

The Mountains of *Chandui.*

Coming from the Seaward, to the Windward of Point St. *Helena*; you'll see this little Mountain over *Chandui*, which bearing E. N. E. at a good Distance, shews thus.
The Hill of *Chandui.*

Coming in to make the Land in the Middle of the Bay of *Guayaquil*, this Hill appears over *Chandui*, and when bearing E. N. E. shews thus.

As you leave this Hill to the *Northward*, it seems to join to the Coast, with this Break and Peek. From

144 *A Description of*

From the Point of St. *Helena*, to the Island *Santa Clara*, lying at the Mouth of the River of *Guayaquil*, is 14 Leagues, and they lie *N. W.* and *S. E.* The Island *Santa Clara* shews differently, as is here represented, and some Ways looks like a dead Body in a Winding-sheet. Ships bound for *Guayaquil* go up to the *Southward* of this Island, steering *E. N. E.* for Point *Arena*, on the Island *Puna*.

The Island *Santa Clara*, in 3 Deg. 20 Min. Latitude *S*.

Bearing from *E.* to *S. E.*

The same bearing from *N. E.* to *N.*

The same bearing *E. N. E.*

From the Island *Santa Clara*, to Point *Arena*, in the Island *Puna*, seven Leagues *E. N. E.* and *W. S. W.* Great Ships cannot pass between the Islands *Puna* and *Santa Clara*, because there are many Shoals; and tho' there be some Channels, none ought to venture in, who are not thoroughly acquainted.

The

The Island *Puna.*

Bearing *West*, distant six Leagues.

The Way from the Island *Santa Clara*, up to *Guayaquil*, is represented in the next Page.

146 *A Description of*
The Town and Port of *Guayaquil*, and Island *Puna*.

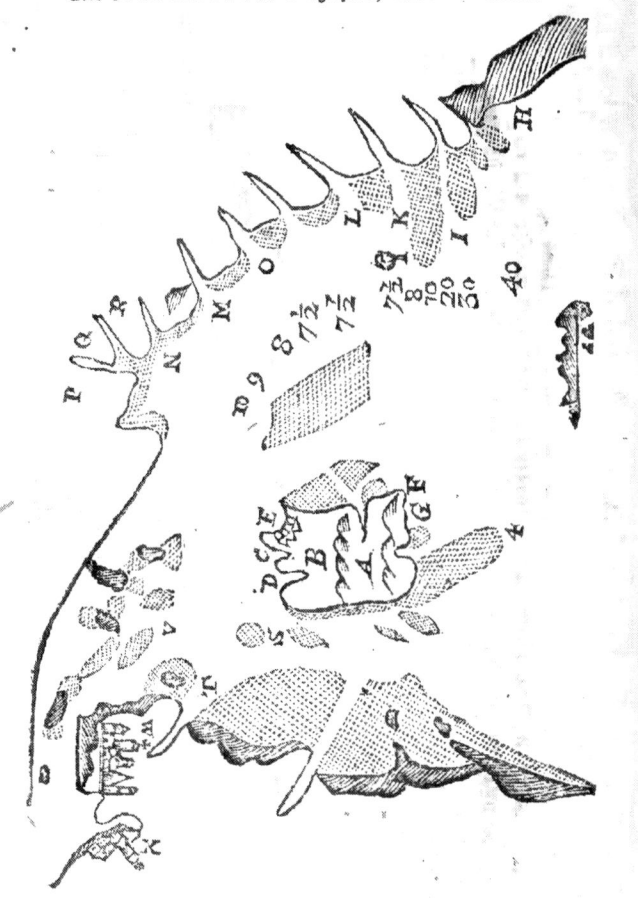

The Letters in this Cut of *Guayaquil* explain'd.

A. The Island *Puna*. B. The high Land of *Mala*. C. The anchoring Ground. D. *Cambra* Creek. E. Point St. *Anne*. F. Point *Arena*. G. The Salt-Pits. H. Cape *Blanco* bearing S.W. I. The River of *Tumbes*. K. *Payana* River. L. Another River *Payana*. M. River *Machala*. N. River *Palao*. O. *Salto del Buey*, or, Ox's Leap. P. River *Suya*. Q. Orange-Garden. R. River *Vola*. S. *Boca Chica*, or, narrow Passage. T. *Isla Verde*, or, Green Island. V. *Bajos de los Frailes*, or, the Fryers Shoals. W. *Guayaquil*. X. The Village call'd *Pueblo de Daule*.

From

The South Sea Coasts.   147

From the Island *Santa Clara*, to *Tumbes*, four Leagues *N.W.* and *S.E.* *Tumbes* is low Land next the Sea, but high up the Country. This Place is known by a large Tree standing at the Mouth of the River, and seen at a great Distance, because it rises above all the other Trees. Come not near this Coast, for there are Shoals running out above a League into the Sea, and it is all flat, because of the Rivers. Two Leagues farther to the Windward, begins a Ridge of high Hills, call'd the high Lands of *Tumbes*, and the Coast runs along to a low Point, call'd the Point of *Mero*. When out at Sea, you'll see these high Lands of *Tumbes*, which have some Breaks, and up the Inland high Mountains.

The Mountain of *Tumbes*.

Standing in from the Seaward, to make the Coast of *Tumbes*, you'll see this Mountain; and when the Break at the Cross bears *S.E.* it shews thus.

From Point *Mero*, to *Cabo Blanco*, or, white Cape, 13 Leagues *N.E.* and *S.W.* the Land next the Sea doubling, that is, shewing one above another; and by the Way there are other high Hills, call'd of *Mancora*; and at the End of them is a little Bay, and a sandy Shore, concluding to the *Southward* in an indifferent high Point. To the Leeward of this Point, and

L 2

148 A Description of

and in the little Bay, Ships may anchor to shelter themselves from the strong Gusts of Wind, and the rouling Sea, which is frequent along this Coast, and particularly at Cape *Blanco*.

From *Tumbes*, to Cape *Blanco*, 14 Leagues, the Land doubling next the Sea, and the Coast all even. About half a League to Leeward of this Cape, is a little Bay, fit to take Shelter in, against the violent Gusts of Wind and high Sea usual at this Cape. All along this Coast, from *Tumbes*, to Cape *Blanco*, the Current always sets upwards; so that there is generally a great Surf, occasion'd by the high Wind and rouling Sea. The best Way to weather this Cape, if there be Squals of Wind, is, to keep close under the Land, where there is good Depth, and make short Trips, under Shelter of the Shore. This Cape is known by its Position, for the Coast downwards lies N. E. and S. W. and that upwards of it N. by W. and S. by E. and at the very Point of the Cape, there is a white Spot at the Edge of the Water.

Cabo Blanco.

Bearing S. W. by S. distant about four Leagues, shews thus.

Bearing N. E. by N. distant six Leagues, thus.

Standing

The South Sea *Coasts.*   149

Standing in to make the Land to the Windward of Cape *Blanco*, this Mountain will appear; and when the Hill at the Crofs bears *Eaſt*, it ſhews thus; and from the Part at (c) the Land runs high, and full of Hillocks to the *Northward*.

From Cape *Blanco*, to Point *Parina*, is ſeven Leagues *North* and *South*, ſome white perpendicular Crags in the Land, ſome ſandy Shores and Bays, which look like Harbours, the chief and greateſt of them about the Mid-way, call'd *Malaca*; and about a League and a half ſhort of *Parina*, is a thick Point, being a perpendicular white Crag, and to the Leeward of it is a good Port, call'd *Fulara*. Only Boats repair to this Port, to load Salt; for there is nothing elſe, nor ſo much as Water. Upon Occaſion you may take Shelter there, if the Current ſets down. They ride here with three Anchors out, one to the *Southward*, another to the S. W. becauſe of the many Squals from Noon 'till Night, and the other to the N. E. on Account of the Land-Breezes. The anchoring is in above 12 Fathom. *Note*, That cloſe to the very Point to Leeward, there is a Shoal, allow for it.

Point *Parina* is low Land, and ſeems to make two little Iſlands. Up the Inland there are high Mountains, and this Point looks as if it run out from the End of the Mountain. At Night, when the Sea runs high, and all is huſh'd, it is ſometimes heard 2 or 3 Leagues off.

L 3   Point

150 *A Description of*

Point *Parina*, in 4 Deg. 20 Min. Latitude *South*.

When bearing *S. E.* diftant three Leagues, fhews thus.

Bearing *North*, diftant three or four Leagues, thus.

From Point *Parina*, to *Paita*, feven Leagues *N. W.* and *S. E.* being a great Bay of low Land, with some white Crags, as far as the River of *Colan*. Venture not into this Bay, for there are often dead Calms, and then again a high Sea, and before the River of *Colan* there are Shoals, which must be kept at a Diftance. From the River of *Colan*, to *Paita*, three Leagues, the Land a little doubling, with white Crags, perpendicular and level at the Top. The Marks to know this Port of *Paita* by, are, a high Hill, with some Breaks, which at firft Sight looks like an Ifland, becaufe the reft of the Land is low. Ships anchor at *Paita* where they can, becaufe fometimes the Wind is large, and fometimes fcant; but the right Part is before the Houfes, in nine or ten Fathom Water.

The

The South Sea *Coasts.* 151

The Saddle of *Paita,* in 5 Deg. Latitude *South.*

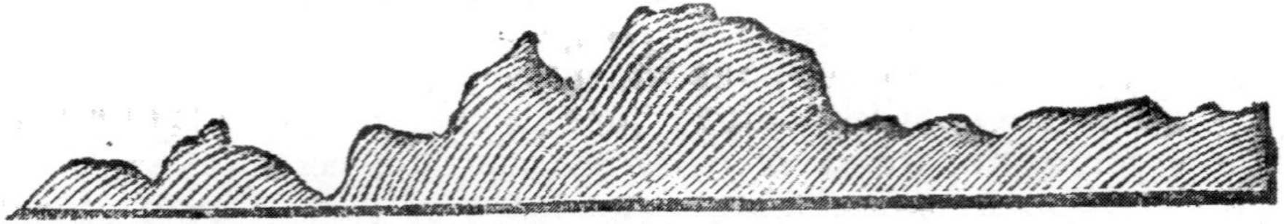

When bearing *S. E.* distant about six Leagues, shews thus.

When bearing from *N. E.* to *N. N. E.* distant eight Leagues, thus.

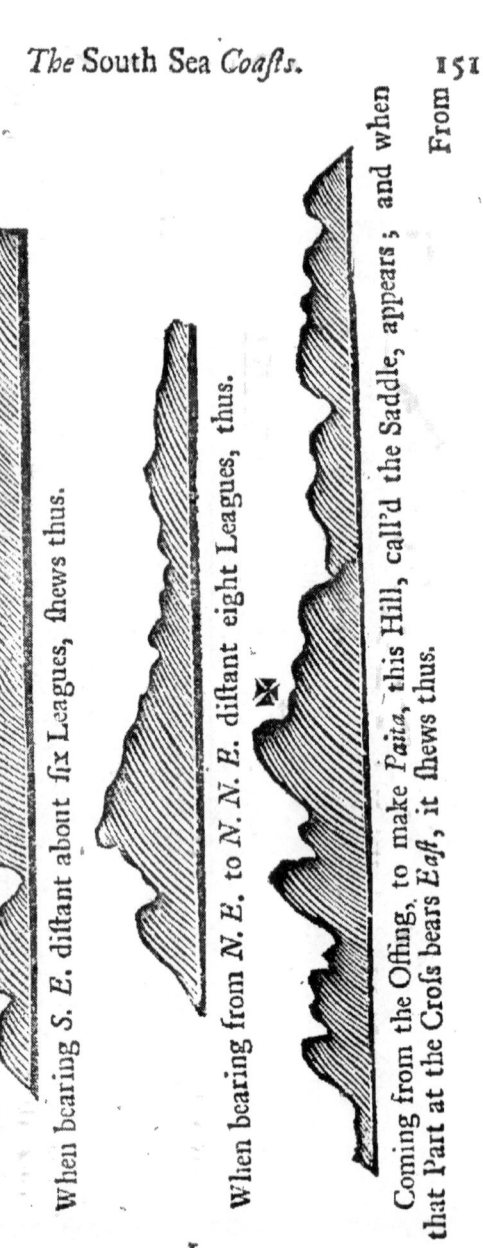

Coming from the Offing, to make *Paita,* this Hill, call'd the Saddle, appears; and when that Part at the Cross bears *East,* it shews thus.

L 4 From

From *Paita*, to *Pena Oradada*, or, the pierc'd Rock, two Leagues, the Land somewhat coping. *Note*, If you are to put into *Paita*, keep an Offing from the Point at the Entrance into the Port; for Ships have been lost on a Shoal there is by it. *Pena Oradada* is a high Rock, which looks whitish from the Seaward, and has a great Hole quite through it.

From this Rock to the Island of *Lobos de Paita*, is two Leagues *North* and *South*. The Coast is not high, but clean and safe, and the Island small and round.

From the Island *Lobos*, or, of Seals, to the Leeward Point of *Aguja*, or, Needle Point, 12 Leagues, forming a large Bay, call'd the Bay of *Cechura*. And from the Island *Lobos*, to *Cechura*, is 12 Leagues; and from the Leeward Point *Aguja*, to *Cechura*, is 10 Leagues *N.E.* and *S.W.* the Coast very low. Ships do not resort to *Cechura*, because there is no lading. All the Bay is flat, but deep and clean Ground; and if there be Occasion, either on Account of Squals, or a rowling Sea, Ships may take Shelter down in this Bay, to the Leeward of Point *Aguja*, where they may also wash and tallow.

Point *Aguja de Sotavento*, or, the Leeward Needle Point, is high white Land; and from this Leeward Point *Aguja*, to the Windward Point *Aguja*, is four Leagues *North* and *South*, high Land, tapering away to the Windward, down to the very Sea. Come not near this Windward Point, for there is generally a great Sea.

The South Sea Coasts. 153

The Windward Point *Aguja*, in 6 Deg. 20 Min. Latitude *South*.

When you come from the Offing, to make Point *Aguja*, it shews thus, bearing from E. N. E. to E. S. E.

Bearing S. S. E. it shews thus.

From this Point *Aguja*, to the Island *Lobos de Sotavento*, or, the Leeward Island of Seals, five Leagues. The Island is about two Leagues in Compass, low, and has some high Rocks and little Breaks. From this, to the Island *Lobos de Barlovento*, or, the Windward Island of Seals, seven Leagues. This is lower than the other, and white; it bears with Point *Aguja* N. and S.

The Windward Island *Lobos*

When bearing from the N. E. to the E. shews thus.

When

154 *A Description of*

When bearing from *W.* to *N.W.* thus.

From the Windward Point *Aguja*, to the Head-land of *Eten*, 19 Leagues, the Coast level with a very sandy Shore, and little Water; but there is frequently a rough Sea, and a Current setting down. There is no high Land to be seen from Point *Aguja*, to the Head-land of *Eten*, but the Head-land it self.

The Head-land of *Eten*

Bearing from the *E.* to *N.E.* shews thus.

The same bearing from *N.E.* to *N.W.*

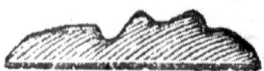

The same bearing from *E.* to *E. S. E.*

The same bearing from *S. E.* to *E.*

The same bearing *E. N. E.* distant eight Leagues.

If you come from the Seaward, to make *Cheripe*, and the Current and Calm should drive you into the Bay, you'll discover over the Head-land of *Eten* another thick lofty Head-land, which may be seen 10 or 12 Leagues off; and if you are much to Leeward, you'll see the Head-land of *Requen*, with a Break on the Top to the *Southward*, making a Peek; and that which lies to the *Northward*, forms a long Table; and within it a little Mountain, up the Inland. As you go off to the *Eastward*, the Mountain will be hid by the Head-land. This Head-land of *Requen* being divided, shews it self in several Shapes. The Head-land of *Eten* appears at the Edge of the Sea like a little Island; it is black, and has a little Break on the Top, the greater Part of it being to the *Southward*.

156   *A Description of*

The Head-land of *Requex*.

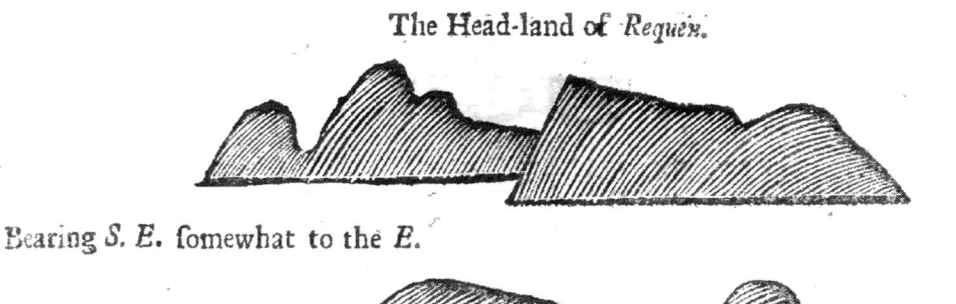

Bearing *S. E.* somewhat to the *E.*

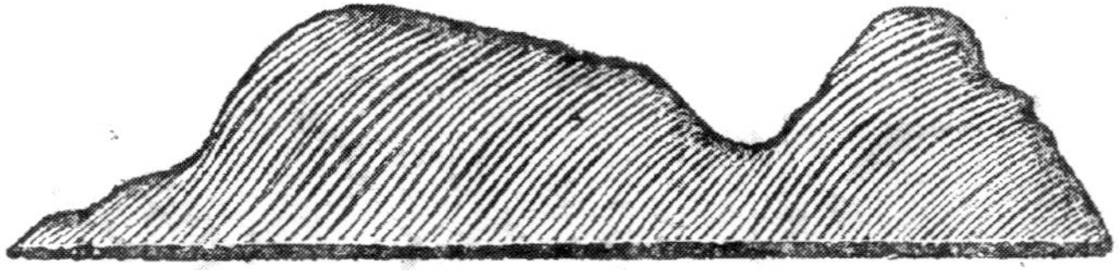

If you come from the Seaward, to make the Island *Lobos*, and cannot see it by Reason of foggy Weather, and go on to the Bay, the Head-land of *Requen*, bearing from *F.* to *N. E.* shews thus.

The

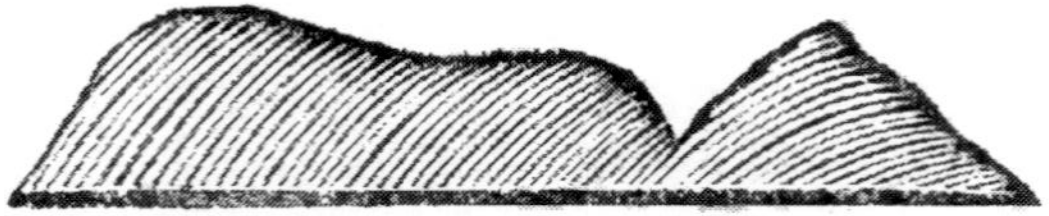

The South Sea Coasts. 157

The same Head-land somewhat to the *Northward* of the *N. E.*

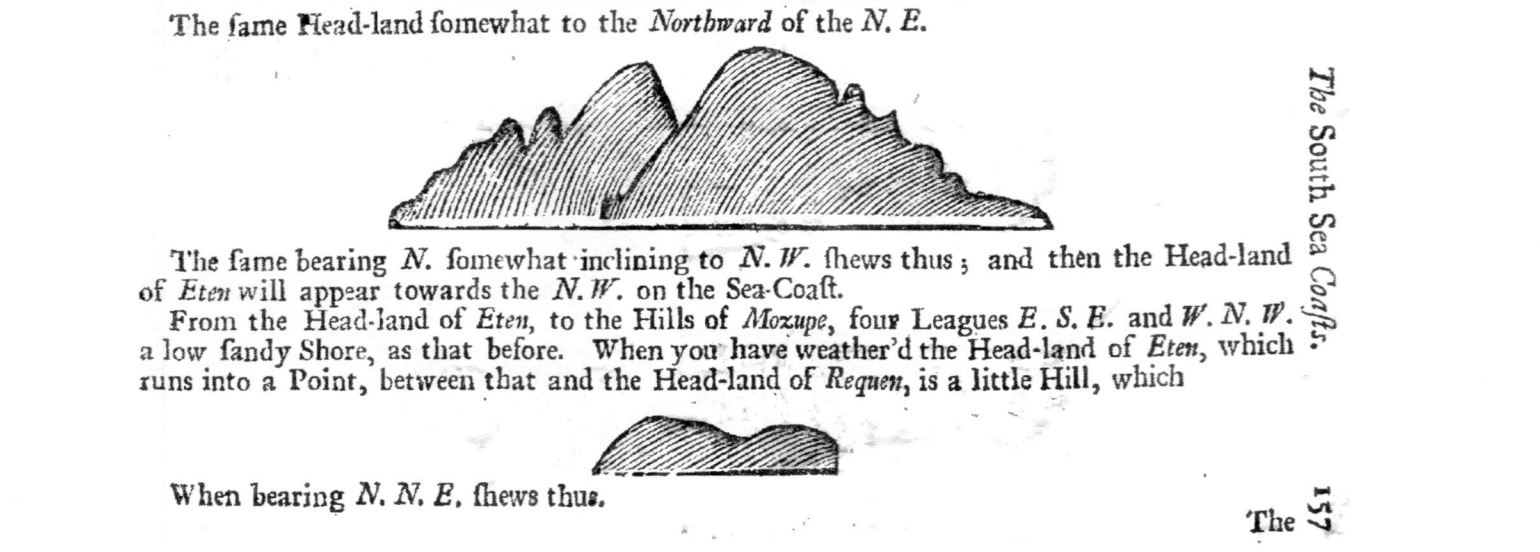

The same Head-land bearing *N.* somewhat inclining to *N. W.* shews thus; and then the Head-land of *Eten* will appear towards the *N.W.* on the Sea-Coast. From the Head-land of *Eten*, to the Hills of *Mozupe*, four Leagues *E. S. E.* and *W. N. W.* a low sandy Shore, as that before. When you have weather'd the Head-land of *Eten*, which runs into a Point, between that and the Head-land of *Requen*, is a little Hill, which

When bearing *N. N. E.* shews thus.

The

A Description of

The Hills of *Mozupe* are black, not very high, and about a League in Length.

The Hills of *Mozupe*.

Bearing *E. N. E.* when you are near Land, the End of them shews thus.

As you leave them, they by Degrees appear round. They are about half a League from the Sea; and between them and the Sea, there are great Pools of fresh Water, which is usually carry'd to the Port of *Cheripe*, a League distant. To the Windward, is a Break of white Sand. You must be very near the Land, to see this Break and Pools, which few observe, because of the Bay. Take Heed not to run into it, for there is usually a great Sea and Calms. Some of the Marks to be observ'd along this Coast, follow.

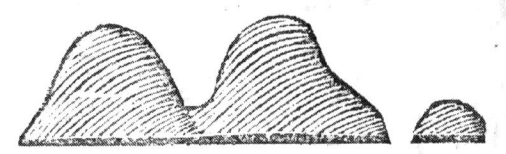

Somewhat to the Leeward of the Head-land of *Eten*, these Hills appear up the Inland.

These Hills to the Leeward of *Cheripe*, bearing *N. by E.* shew thus.

The South Sea Coasts. 159

The Mountains of Eten.

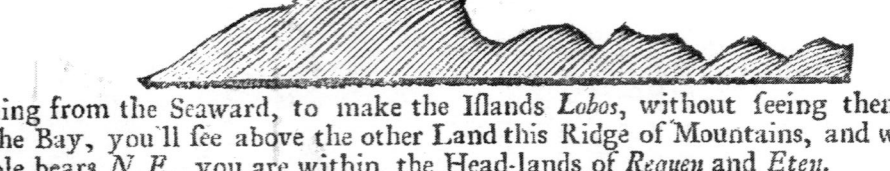

Coming from the Seaward, to make the Islands *Lobos*, without seeing them, and passing on to the Bay, you'll see above the other Land this Ridge of Mountains, and when the highest Table bears *N. E.* you are within the Head-lands of *Requen* and *Eten*.

From the above Break, or Creek of white Sand, to the Port of *Cheripe*, two Leagues, higher Land than that above, lying *North* and *South*, with some red Crags. To the Leeward of *Cheripe*, is a Hill half a League up the Inland, higher and longer than the Hills of *Mozupe*, which shews severally, according to the different Positions. At a Distance it looks like an Island, and the Head-land here mark'd with a Cross, like a high Rock in the Sea.

The Hill to the Leeward of *Cheripe*

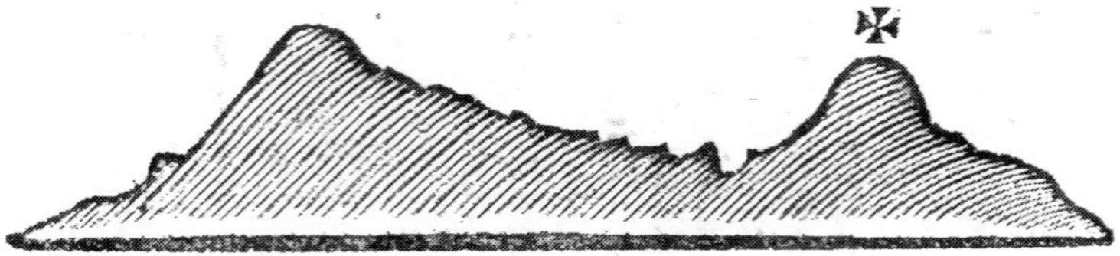

Bearing *East*, shews thus.

160   *A* Description *of*

St. P**e**ter's Hills.

When you make the Land to Windward of *Cheripe*, you'll see these which are call'd St. *Peter*'s Hills. When the Hill at the Cross bears *East*, it shews as there represented; but if you leave it to the *S. E.* a Passage opens where the (o) is, and the two Hills divide.

Making the Land above the Port of *Cheripe*, somewhat to Leeward; and this Heed-land, which is also to Leeward of *Cheripe*, bearing *E.* it shews thus; and at a great Distance it looks like *two* Islands, parting where the Cross is.

The South Sea Coasts.

Guadalupe Sugar-Loaf.

Over *Cheripe* appears this Hill, call'd *Pan de azucor de Guadalupe*, or *Guadalupe Sugar-loaf*, which bearing from *E.* to *S.E.* shews thus.

If you would anchor in the Port of *Cheripe*, which is the Port of the *Vaies*, the Place is to the Leeward, under a low Point, which at a Distance looks like a little black Island; and if the Weather be clear, you'll see the white Church at above three Leagues Distance. In coming into this Port, take Heed of the Windward Point, which has a very dangerous Shoal, with a Ridge of Rocks running above half a League into the Sea. Sound all the Way, keeping in eight Fathom Water, and giving the Shoal a Berth; then make directly for the Church, and when it bears *F.S.E.* and a great Cross you'll see there at the Corner of the Church *South*, and you have between seven and eight Fathom Water, drop your Anchor, for there is the Port.

From Port *Cheripe*, which is in seven Degrees of *South* Latitude, to *Pacasmayo*, six Leagues *N.W.* and *S.E.* low Land and Sands, with some Crags here and there; and about half a League up the Inland, some Hills, call'd of *St. Peter de Illoque*; and at the *North End*, is a round

162 *A Description of*

round Hill, call'd the Sugar-loaf of *Guadalupe*, which has a little Break at the Top when bearing *East*. These Hills of St. *Peter* and *Guadalupe* Sugar-loaf represented before.

Pacasmayo makes a great Strand on the Shore, and in the Midst of it a high Rook. To the Leeward of this Rock, which stands on the Land, there is 12 Fathom Water. Take Notice, that this is a foul dangerous Coast, and Ships do not go to load in this Port, because of the high Sea, and Hazard of being lost.

From *Pacasmayo*, to *Malabrigo*, five Leagues, low Land and sandy, with some low whitish Crags. About three Leagues short of *Malabrigo*, begins a Bay enclos'd with low Sands, and reaching to the Port of *Malabrigo*, which is a Harbour for the Vales. To com to an Anchor in it, you must found all the Way, in five or six Fathom Water, keeping close up with the Windward Head-land, 'till you come into four Fathom and a half; and when the high Break in the Head-land bears *South*, let fall your Anchor.

There is frequently a great Sea in this Port of *Malabrigo*, and it blows hard; and if you come from the Seaward, you'll see a Mountain, which at the End to the *Southward* has many small Breaks, and to the *Northward* on the Top of the Mountain forms a round Peek, not very high.

This

The South Sea Coasts. 163

This Ridge of Hills appears at a great Distance in clear Weather, because it is high, and is seen from the Leeward of *Cheripe*, as far as to the Windward of *Malabrigo*. When the Peek at the Cross bears *East*, you are off *Cheripe*; and when *N. E.* off *Malabrigo*.
The Hill of *Malabrigo*.

When this Hill, which is above the Port of *Malabrigo*, bears *E. S. E.* it shews thus; and over it you'll see this next below.

This Hill, bearing *East*, shews thus.

M 2

The Head-land of *Malabrigo*, bearing *E. S. E.* distant between three and four Leagues, shews thus.

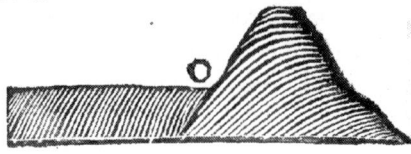

The high Rock of *Malabrigo*, from the *East* to the *North*, shews thus, and open where the (o) is, when left to the *S. E.*

From *Malabrigo*, to *Guanchaco*, which is the Port to the City of *Truxillo*, 11 Leagues. In the Mid-way is a great River, call'd the *Magdalen of Cao*; the Coast low and sandy, with many large Hills, and little Mountains, divided from one another, a little up the Country. Two Leagues short of *Guanchaco*, there rises a Point, and runs along somewhat higher than the other Land before, as far as *Guanchaco*. If you would anchor there, you must sound all the Way; and when the Church, which you will see look white, appears over the Town, and the Bell Hill bears *N. E.* by *N.* drop your Anchor in 10 Fathom Water. There is generally a great Sea in this Port. Take Heed often to observe your Anchor and Cable; for the Anchors sink much, by Reason of the great Sea.

From the Port of *Guanchaco*, to the Head-land of *Guanape*, nine Leagues *N. N. W.* and *S. S. E.* It forms a Bay half Way, at the Bottom of which is the Headland of *Colletas*; the Coast is foul, and low next the Sea. None anchor at this Head-land of *Guanape*, unless in Case of absolute Necessity, because the Coast is, as has been said, foul, and there is a great Sea.

The

The Head-land of *Guanape*, in 8 Deg. 30 Min. *South* Latitude.

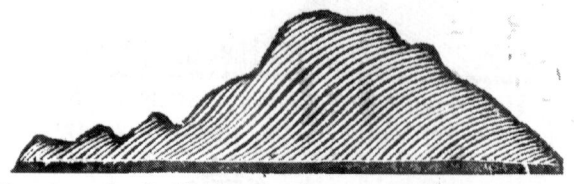

Bearing *S. S. W.* shews thus; and at a great Distance looks like an Island, because the other Land is low.

The same bearing *N. E.* shews thus.

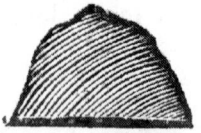

The high Rock bearing *North*, thus.

Without the Head-land of *Guanape*, is a large high Rock, call'd *Farellon de Guanape*; and to the Landward of it, a little Island somewhat lower. Ships may pass on any Side of this high Rock, for there is every where Water enough, and the Ground is clean.

Farellon de Guanape, in 8 Deg. 30 Min. *South* Latitude.

When it bears *N. E.* shews thus.

166 A Description of

The high Rock of *Malabrigo*, and that of *Guanape*, bear from one another *N. N.W.* and *S.S.E.* There is good anchoring to the Leeward of this Rock of *Guanape*, in a sandy Creek, in six Fathom Water, two Musket Shot from the Shore. If the Boat goes up for Water, it must be at the Flood, because the River has rais'd a Bank, where Boats stick at the Ebb. Formerly Ships us'd to lade at this Port of *Guanape*, for *Panama*.

From the Head-land of *Guanape*, to that of *Chao*, five Leagues, *N.W.* and *S. E.* the Coast low next the Sea; but the Head-land of *Chao* is high, and has three or four little white high Rocks standing up before it; and to the Windward of it, is a little blackish Island. There is no anchoring at this Cape, because all the Coast is boisterous. The little Island before the Bay of *Chao*, bears with the Head-land of *Guanape N.W.* and *S. E.* Short of that Island, there are dangerous Shoals. In the Bay, is a Port well shelter'd against the *South* Wind, but no fresh Water, unless brought from the Town, a League up the Country.

The Island of *Chao*.

The

The South Sea Coasts. 167

The Head-land of *Chao*.

Standing in from the Seaward, to make the Head-land of *Chao*, when it bears *East*, it shews thus, and the Head-land of *Guanape* may be seen to the Leeward.

To the Windward of *Farellon de Malabrigo*.

Coming from the Seaward, to make the Rock of *Malabrigo* to the Windward, this Mountain appears; and when the Part at the Cross bears *East*, shews thus, and that same Hill is the Bell of *Truxillo*.

M 4

A Description of

Mountain to Leeward of *Truxillo*.

In the Offing appears this Mountain to the Leeward of *Truxillo*, as far as near *Santa*, and the Peeks at the Cross are over *Truxillo*; and when it bears E. N. E. you are to Windward of *Truxillo*.

The Bell of *Truxillo*.

When this Bell of *Truxillo* bears N. E. distant four or five Leagues, it shews thus; for at a greater Distance it looks otherwise, and to the S. E. is a Sugar-loaf Hill.

Between

The South Sea Coasts.

Between Guanchaco and Guanape.

This Hill appears above the Head-land of *Caretas*; and when bearing *East*, shews thus. At the same Time the Head-land of *Guanape* appears to the Windward.

The Bell of *Truxillo*.

Bearing *N. N. W.* at a great Distance, shews thus.

The Bell of *Truxillo*, bearing *N. by E.* shews thus.

A Description of

From the Head-land of *Chao*, to the Port of *Santa*, four Leagues *N. W.* and *S. E.* the Coast low, with some high Hills. Short of *Santa*, is a high Rock, call'd *el Corcobado*; pass not between it and the Land, for there are Shoals; and before the Port of *Santa*, you'll see an Island, about a League in Length, lying *North* and *South*. You may pass by either End of this Island into the Port, for it is all deep Water, the Passage being to the Leeward of the Island into the Port, when coming from below; and between the Island and the Continent, when from above. The anchoring Place is at the Foot of a Hill, that is on the Continent, where is a little Bay, lying *East* and *West*, in eight Fathom Water, beyond the first Point of the Head-land, where there are some sandy Creeks, for about a Stones Throw.

The Head-land of *Santa*.

When the Head-land at the Cross bears *S. S. E.* distant five Leagues, it shews thus.

The Island of *Santa*, in bare 9 Degrees *South* Latitude,

Bearing *North*, distant four Leagues, shews thus.

Chinbote.

Chinbote.

When you come from the Windward, to make *Santa*, this Hill of *Chinbote*, at the Cross, shews thus.

From the Windward Point of the Island *Santa*, to *Ferol*, one League; and from the same Point, to *Casma*, 10 Leagues *N.W.* by *N.* and *S.E.* by *S.* high Land, full of Hillocks. The Port of *Ferol* is good and safe, the Passage into it between some small Islands, lying before the Entrance, where there is nothing to fear, for it is all clean and safe. There is a great Hill on each Side of this Port of *Ferol*, and the little Islands are between those Hills. The Hill to the *Southward*, is round and large, and has some great Spots in the Middle, towards the Sea. Before this Hill is a Shoal, near the Continent.

A Description of

The Channels into the Port of Ferol, in 9 Deg. 40 Min. South Latitude.

From the Port of Ferol, to Bombacho, four Leagues. This is a good Port, tho' not frequented, becaufe there is no Trade. There is nothing to fear, but what may be feen. If you would anchor there, take Notice, that to the Windward of the Head-land is a fmall Rock above Water; keep clofe to it, and go on to anchor at the faid Head-land in five or fix Fathom Water. There is no Shoal; and when the Boat would go afhore, it muft be near the River, leaving it on the left Hand. The going out is bad, becaufe it is to the Weftward, to weather the Leeward Point. You muft go out in the Afternoon, before the Wind falls; for it grows calm every Night in this Bay. When you weigh, all the Sails muft be loofe, and veer'd out at once, the Bowlings clofe hal'd; for you muft go upon a Wind. Here is Wood, Water, and Flefh. The Place for the Boat to take in Lading, is a fmall fandy Shore, with a high Rock in the Midft of it, thro' which there is a Hole. Bombacho is in 9 Deg. 30 Min. of South Latitude.

Between

The South Sea Coasts.

Between *Ferol* and *Bonbacho*.

Coming from the Seaward, to make *Bonbacho*, when that Port at the Crofs bears *East*, the Land shews thus.

From *Banbacho*, to *Casma*, four Leagues; and by the Way there are some small Islands, and very deep Bays, which you'll not see from the Offing, because close in with the Continent. *Casma* is a good Port, tho' the Wind blows hard from Noon forward; however, it makes no Sea, and at Night is quite calm. Without in the Bay is a little round low Rock, level with the Water, somewhat more inclining to the *Northward*, than to the *Southward*, and it has a Shoal on the *South* side, before Mount *Calvary*, under Water, and not to be seen but when the Sea is low, and then it has a Fathom and a half Water. Ships may pass between this Shoal and the Continent, without any Danger, for there is 14 Fathom Water; and coming in large, they may cling as close as they will to the Continent. The anchoring is at a white Head-land. They may careen at the same Head-land, for there is Conveniency to bring the Ships down.

174 A Description of

Cafma, in 10 Degrees of *South* Latitude.

Coming from the Sea-ward, to make *Mongon*, to the Leeward of it, you'll see this Land, which is that of *Cafma*; and if *Mongon* does not appear, as being often cover'd with Fogs, the Point at the Crofs looks like little Iflands; and where the upper Crofs is, the Land rifes to the S. E. as far as *Mongon*.

When this Point bears from N.E. to N.N.W. it fhews thus. You will not fee the two Hillocks at the Crofs at firft; but as you come nearer, they will look like Rocks in the Sea; and foon after you'll fee they are Part of the Continent.

From *Cafma*, to *Mongon*, three Leagues; and along this Coaft there is generally a Current fetting down, occafion'd by the Shore and Point. *Mongon* is the higheft Hill on that Coaft, feen at a great Diftance, having a little Break on the Top, and fhews in feveral Shapes. Bearing S. E. it appears plain at the Top, like a Table; from E. S. E. to E. N. E. it looks round, with the little Break on the Top; and as you leave it towards the N. E. the Break opens more, and grows longer. The Current fetting here, as was faid above, downwards, often ftops Ships in their Paffage.

Mongon,

The South Sea Coasts.

Mongon,

Bearing from *E.* to *N.E.* at a great Distance, shews thus; and seen somewhat nearer, makes another little Head at the *N.W.* Point.

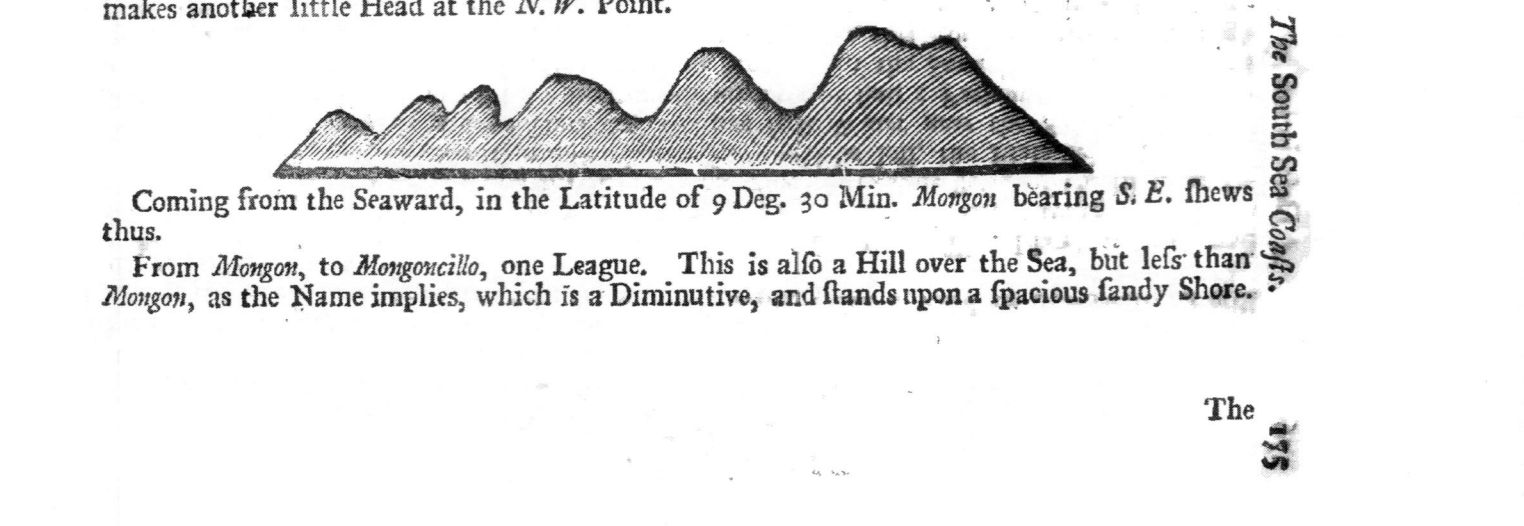

Coming from the Seaward, in the Latitude of 9 Deg. 30 Min. *Mongon* bearing *S.E.* shews thus.

From *Mongon,* to *Mongoncillo,* one League. This is also a Hill over the Sea, but less than *Mongon,* as the Name implies, which is a Diminutive, and stands upon a spacious sandy Shore.

The

Bearing E. N. E. shews thus; and if you coast along either upward or downward, it will grow round by Degrees.

From *Mongon*, to the Island of *Puerto Vermejo*, four Leagues. This Island is small and white, and in the Way to it is a Bay, call'd el *Jaguey de las Culebras*, having two Points running out, like a Harbour; and here begin some Flats, a League short of the Island of *Puerto Vermejo*. Go not between it and the Land, for you'll be loft, the Flats being very dangerous. *Puerto Vermejo* is a good Harbour, and has fresh Water near the Sea. At a small Distance from the Shore, you'll see a little Well, with about half a Fathom Water in it; and wheresoever you go but 10 or 12 Paces from the Sea, dig in the Sand half a Fathom, and you'll find reasonable good Water, not very brackish. Over this Port is a thick and high Hill, the Slit whereof runs *Southward* down to the Sea, and on the *North* Side it is almost perpendicular.

Puerto Bermejo.

Coming up to Windward of *Puerto Bermejo*, you'll see this Hill, which, bearing *East*, shews thus, either near or at a Distance; and at the Head-land, (o) down by the Sea, it is all white Sloughs; and that which is to Windward of all the rest, looks like a half Moon, with the Horns upward; and within it is another white Spot, and without you'll see something white, like two Sails.

From *Puerto Bermejo*, to the Port of *Guarmei*, three Leagues, low Land next the Sea. This Port of *Guarmei* has a low level Point on the *South* Side of it; and a little Way up the Inland, there are many Hillocks and high Ridges; and over the Port somewhat up the Country, there are two round Hills, thicker and higher than any of the others. That which is to the *Northward*, bigger than the other to the *Southward*. The anchoring Place is to the Leeward of the low Point above-mention'd, in seven Fathom Water, keeping a Rock you'll there see above Water right astern. The Anchors are sometimes apt to drag, with the violent Wind, and therefore it is best to go farther in, for it is all safe, and you may anchor in six Fathom Water. Here you may water, for the River runs down near the Landing-Place, where is an *Indian* Town, and you may be furnish'd with any Provisions you want.

Guarmei,

A Description of

Guarnei, in 10 Deg. 30 Min. of *South* Latitude.

When you make *Guarnei* from the Leeward of the Port, this Hill bearing *East*, shews thus.

The same Hill bearing E. by N. shews thus.

From the N.E. to the N. distant three Leagues, it shews thus, and the high Land runs out to the Leeward; and to the *Eastward* is a high Hill, with three Breaks on the Top.

Bearing N.E. distant four Leagues, it shews thus.

The

The Dugs of *Puerto Bermejo*.

Between *Puerto Bermejo* and *Guarmei*, is a Hill, which looks like two Dugs, and is therefore call'd *Las Tetas de Puerto Bermejo*, that is, the Dugs of *Puerto Bermejo*, or *Red Port*, and coming up from the Offing directly upon it, shews thus.

From the Port of *Guarmei*, to *Jaguei de la Zorra* four Leagues, the Coast next the Sea low. *Jaguei* is a double Hill; and if you come towards it from the Offing, there appears a little Break on the Top, leaving the greater Part of the Hill to the *Northward*, and falls away on that Side quite down to the Sea. And a little more to the *Southward*, is an indifferent Sort of a Hill, which, if seen when there is a Fog on the Coast, looks like the Rock of *Maltesi*, or *Marsoque*.

The high Land of *Jaguei de la Zorra*,

Bearing *South*, distant five or six Leagues, shews thus.

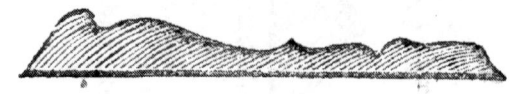

Bearing *N. N. E.* distant four Leagues, it shews thus.

180   A Description of

The same bearing *East*, shews thus.

From *Jaguei de la Zorra*, or, as some call it, *Hoguei de la Costa*, to the River *de la Barranca*, nine Leagues, most low Land, with Hillocks next the Coast, and at the half Way a thick high Hill, call'd *Cerro del Gramadal*, on the Top whereof are two Breaks, and the middle Hill higher and bigger than the others; that to the *Southward* next in Bigness, and the other to the *Northward* the least, and roundish. *Note*, That if the Land be foggy, the Hills on this Coast look like Islands.

Barranca.

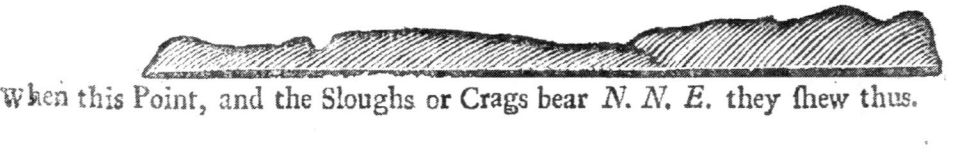

When this Point, and the Sloughs or Crags bear *N. N. E.* they shew thus.

The

The South Sea Coasts.

The Break of *Barranea*.

When you come from the Offing, to make *Barranca*, somewhat to Windward, you'll see the Break at (o) which bearing E.S.E. shews thus.

When the Break of *Barranca* bears E.N.E. it shews as you see at the Peek where the (o) is.

When the Break bears from E. to N.E. distant about six Leagues, it shews thus; and over the Hill at the Cross there sometimes appear Hills cover'd with Snow.

Jagui de la Zorra, or, *Haguei de la Costa*.

Making the Land as far on as *Barranca*, it the Ridge of Mountains happens to be cover'd by a Fog, and cannot be seen, looking towards the *N. E.* somewhat *Northerly*, you'll see this Ridge, which is *Haguei de la Costa*.

A League short of the River *Barranca*, is *Paramonguilla*, or *Paramonga*, being a small Ridge, or Hill, but looks like a white Rock in the Sea; and in the Offing looks white, like a Ship under Sail.

Paramonguilla,

Bearing *S. E.* shews thus.

Bearing *N. E.* it shews thus.

To the Leeward of *Paramonguilla*, is a low Point, but perpendicular and black; and to the Leeward of that, is a sandy Shore; where Ships may anchor in six or seven Fathom Water, in case of Necessity, either for the Current setting down, or Squals of Wind.

The Hill *Gramadal*,

Said above to be half Way between *Huguei* and *Barranca*, bearing *N. E.* at a considerable Distance, shews thus.

The River *de la Barranca*.

When you are come the Length of the *Barranca*, you'll see this Head-land; which, to the Windward of the Point at the Cross, shews thus, and there runs out the *River de la Barranca*.

A Description of

The Ridge to the Windward of the River *de la Barranca.*

You'll see this Head-land to the Windward of the River *de la Barranca*; and when the Peek at the Crofs bears *N. E.* it shews thus, and the River falls into the Sea to the Leeward of the Head-land at (o).

When this Break at *Barranca* bears *N. E.* somewhat to Leeward, the Land at (o) shews thus.

From the River *de la Barranca*, to *Supe*, two Leagues; and to the Leeward of the Strand of *Supe*, there are some red Sloughs or Crags next the Sea; and to the Leeward of them again is a little low Point; and to the Leeward of that Point, is the Port of *Barranca*, where you anchor in seven Fathom Water, and muſt moor, that the Ship may not come about, becauſe of the Land-Breeze. The Bearings along all the Coaſt hither, are above. *Supe* is a Bay forming a ſpacious ſandy Strand. Here trading Boats uſe to come to an anchor, to load Corn. Take Notice there is often a great Sea in it; if you go aſhore with the Boat, remember the Sea breaks on it.

From

The South Sea *Coasts.* 185

From this Strand of *Supe*, to the Island of *Don Martin*, two Leagues, the Land low next the Sea, but a little up the Country there are some high Ridges, some of them like little burning Mountains, one of them the next represented here below. This Island of *Don Martin* is white, about a quarter of a League from the Continent, and its Compass is about half a League. It is plain, and not very high. When this Hill bears *East*, you are then the Length of the Island.

The Hill of *Supe*.

Bearing *East*, shews thus; but as you leave it towards the *N. E.* it will, by Degrees, look level at the Top, only where the Cross is will be a little Break, and the Peek at (o) becomes round, like a little burning Mountain.

Mountains to the Windward of *Supe*.

When this Point, which is to Windward of *Supe*, bears *N. E.* distant two Leagues, it shews thus.

From

186 A Description of

From the Island of *Don Martin*, to the Port of *Guaura*, one League; and as soon as pass'd the Island to Windward, there is another small Island, call'd *de Lobos*, that is, of Seals, which has a Shoal close by it. Pass not between the two Islands, nor between them and the Continent, for there is but little Water. If you would anchor in this Port of *Guaura*, keep the little Island *Lobos* right a-stern. On the Top of the Head-land there are two Pieces of Wall, which look like two Pillars; when you have brought them together, and the little Island is right a-stern, you may anchor, for to the Windward there are Mice. There is generally a great Sea in this Port, the Town is a League from it; but there is Water, and all other Necessaries.

Guaura.

When you are up with the Length of *Guaura*, and the Land at the Cross bears *East*, it shews thus, and the Head-land of *Salinas* will be to the S. E.

The Head-land of *Guaura*,

Bearing *N. E.* shews thus.

From

From the Port of *Guaura*, to *Guacho*, one League; and here some of the trading Boats anchor to the Leeward of a little Head-land you'll see. Come not too near it on the Windward Side, for it has a dangerous Shoal, on which the Sea uses to break.

From Port *Guaura*, to *Salinas*, or, the Salt-Pits, two Leagues, low Land next the Sea. The Port is safe, tho' subject to much Wind, and a great Sea, but has neither Wood nor Water; so that the Ships which go to the Salt-Pits, if they want any Thing, send for it to *Guaura*, either by Sea or Land. The anchoring Place here, is before you come to the End of the Salt-Pits, in six or seven Fathom Water, where the sandy Shore begins, and is call'd the Port *de la Barca*. This Port has a Head-land, which throws out two Skirts towards the Sea, that to the Leeward the smallest; and when near, there appears a little Break on the Top of it, where the Cross is; and as you leave it towards the *S. E.* and to the Leeward, that Break opens, 'till it appears plainer.

The Head-land of *Salinas*, in 11 Deg. 30 Min. *South* Latitude,

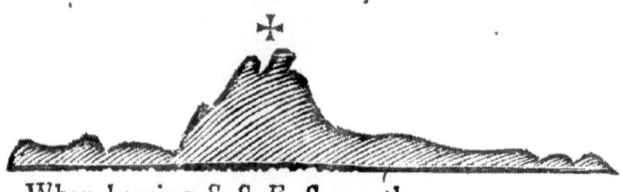

When bearing *S. S. E.* shews thus.

The same bearing from *N. E.* to *N. W.* at a good Distance, shews thus.

From the Head-land of *Salinas*, to *Maltesi*, which is farthest in the Offing of all the abovesaid Rocks

of

of *Guaura*, otherwise call'd *Chontales*, four Leagues N. E. and S. W. and from *Maltesi*, to another great above Sea Rock, call'd *Marsoque*, a League and a half; and from *Marsoque*, to the Continent, about two Leagues. There are seven or eight of these above Sea Rocks of *Guaura*, between great and small, and they bear from one another E. N. E. and W. S. W. All about them is clean and deep Water, as well as without; and if you will pass between *Maltesi* and *Marsoque*, you may do it safely, for there is above 40 Fathom Water; but when you would pass between them, keep somewhat to Windward, and let your Anchors be ready to drop, if there be Occasion. About these Rocks of *Guaura*, there is frequently a strong Current downwards, and Squals of Wind; and such Ships as cannot weather them, resort to *Salinas* for Water, and what else they stand in need of. This Rock of *Maltesi*, and the Island of *Don Martin*, bear to one another N. E. by N. and S. W. by S. and *Maltesi* and the *Hormigos* bear to one another North and South, distant seven Leagues. *Maltesi*, and the Island of *Callao*, bear to one another N. W. by N. and S. W. by S. distant fifteen Leagues. All the Coast from below, that is, from *Santa*, to these above Sea Rocks, has deep Water, and is clean. This Rock of *Maltesi* is in the Latitude of 11 Deg. 40 Min. South. *Maltesi* Rock is two Leagues and a half from the Continent; and *Marsoque*, which is nearer the Land, is larger, and they are above a League distant from each other.

Maltesi. *Marsoque.* The other Rocks.

These above Sea Rocks, near about *South*, shew thus.

From *Maltesi*, to the Strand, and *Tanbo*, that is, the Inn, *de las Perdices*, or, of Partridges, five Leagues
East

East and *West*. When you come from the Offing, to make these above Sea Rocks, take Heed not to run into the Bay of *Chancay*, when you discover them; for there are commonly dead Calms, and a great Sea, and therefore stand away presently for the *Hormigas*.

From the Rock of *Maltesi*, the farthest to Seaward of those of *Guaura*, to *Hormigas*, seven Leagues *North* and *South*. This Island of *Hormigas* is small and white, and has a little Break in the Middle of the *South* side, the Bottom good and clean; and on the *North* and *N. W.* there is a Flat of Rocks, stretching out above a League, and at the End of that Flat is a little above Sea Rock; by Day you'll see the Sea break on it, and by Night you may hear the Sea roar above a League to the Offing. Take Heed how you pass by this *Hormigas*, for Ships have been cast away there. If you are come the Length of, and would make it, and cannot see it by Day, tho' the Wind be large, do not make too long a Trip, but ply off and on all the Night, and make no long Runs towards the Land; for sometimes it may happen to be calm, and then you'll not hear the breaking of the Sea, but may be a-ground, as has happen'd to some who took too long a Trip towards the Land. *Hormigas* and *Maltesi* bear to one another *North* and *South*, distant seven Leagues. *Hormigas* and the Island of *Callao W.* by *N.* and *E.* by *S.* distant eight Leagues. *Hormigas* and *los Pescadores E.* by *N.* and *W.* by *S.* distant nine Leagues.

To return to the Continent, and the Coast running from *Salinas*, or, the Salt-Pits, to the Windward. *Note*, That near the Point, and End of the Coast, opposite to the above Sea Rocks of *Guaura*, there is a little Bay, call'd *la Herradura*, or, the Horse-Shoe. This is a good Port, and sometimes Ships which cannot weather the aforesaid above Sea Rocks, put in for Shelter between this Point of the Continent, and a little above Sea Rock, call'd *el Tambillo*, in their Way downward; but it is best to keep without this Point the
Land

Land makes, which comes from below, and proceed to *Tanbo* and *Playa de las Perdices*, or, the Strand of Partridges, which is three Leagues, low Land, but a little up the Inland is a high sandy Ridge.

At this *Playa de las Perdices* there is good anchoring, and a clean Bottom, being deep Sloughs of Sand, and then a high Ridge which falls away to the *N.W.*

From *Playa de las Perdices*, to *Chancaillo*, three Leagues.

Chancaillo.

Over the Road of *Chancaillo* appears this Hill, which shews thus, whatsoever Way it is seen, but most exactly bearing *N.E.*

From *Chancaillo*, to *Chancai*, two Leagues, all Sloughs next the Sea, which at a Distance look black, and are green Plats of Grass and Sedge in the Sloughs, occasion'd by the Plenty of Water running down them to the Sea from all the Marshes. If you are near Land, you'll see some little Streams of Water fall into the Sea; as also the Town, and the white Walls of the Church of St. *Francis*. *Chancai* is a good Port, land-lock'd against the *South* Wind; but it has sometimes a great Sea, because of the Sea Breezes, which last long, blow directly in, and there is no Shelter against them. To anchor here, run along close by the Head-land, for the Port is to Leeward of it; and there you may anchor where you will, for it is all clean, provided you come not too near the little Bay the Port makes, for there are Mice within; and close under the very Head-land there is eight Fathom Water, clean Ground. Here you may have Water and Provisions. Over the Town is a Hill.

The

The Hill of *Chancai*, behind the Town.

When bearing N. E. shews thus.

Run not into this Bay of *Chancai*, for there are frequently dead Calms, and a great Sea, which will drive you upon the Coast; and therefore, as I said above, rather chuse to go by the Island *Hormigas*. If you are bound for this Port of *Chancai*, you will know it by the Head-land underneath, for the Port is to the Leeward of it.

The Head-land of *Chancai*,

Bearing N. E. shews thus.

From the Port of *Chancai*, to the great above Sea Rock, call'd *de los Pescadores*, or, of the Fisher-men, three Leagues, high Land next the Sea, and makes a Break in the Middle. This high Land is call'd *el Cerro de la Arena*, that is, the Sand-Hill. *East* and *West* with the largest of the Rocks call'd *Pescadores*, is a Port, which some call *Puerto del Ancon de Rodas*, and others *el Ancon*. It is a good safe Port, a League distant from the aforesaid *Pescador* Rock; the Passage into it, is to the N. W. of the great Rock, for it is all clean. There is some Water to be had in some small Wells, but a little brackish, and no Sea.

These Rocks *de los Pescadores*, are six in Number, great and small, all white, and that to the N. W. is the biggest. They bear from one another E. N. E. and W. S. W. the great one and *Hormigas* bear from one another E. by N. and W. by S. and from the Island of *Callao* N. N. E. and S. S. W.

The

A Description of

The Island of *Callao*, in 3 Deg. 20 Min. Latitude South.

When you are near the Coast of *Chancai*, and the *Pescadores*, bearing near about *South*, shews thus.

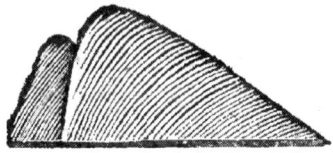

The same Island bearing *S. E.* distant four Leagues, shews thus.

The same bearing *E. S. E.* shews thus; and as you leave it to the *East*, the Break at the Cross opens by Degrees.

The same bearing *E.* by *N.* shews thus.

The

The same when you are to the Windward, and it bears *N.N.W.* six or seven Leagues distant, shews thus.

When you come from the Seaward, to make this Island, and discover it at a great Distance bearing *N.E.* it shews thus; and the Breaks at the Cross by Degrees seem to sink even with the Water.

This Island is about two Leagues in Compass, and at the End of it has some very small Islands and above Sea Rocks; and beyond them another little Island, high and perpendicular to the Sea; and no Ships can pass between them, because the Distance is small, and but little Water in it.

From the *Pescadores* Rocks, to the Port of *Callao*, is five Leagues *N.N.W.* and *S.S.E.* and about the same Distance to the Head-land of the Island. From the *Pescadores*, to the high Rock call'd *de Dona Francisca*, the Land is high, and from thence to *Callao* low. In this Bay, which is form'd between the Island of *Callao* and the *Pescadores*, Ships may ply upon a Wind, and

and anchor any where, as they may all along the Coast from *Chancaillo*, for it is all clean and deep. The several Bearings of this Island are sufficiently expressed above. If you would pass in to Leeward of it, give it a Berth of at least a League, because it throws out many Sands; and sometimes there are such Gusts, that Ships cannot weather the Head-land in half a Day. When in, you may ply up and down, and anchor any where, for it is all clean and safe, in 10 or 12 Fathom Water; and even down to four Fathom there is no Danger. The Head-land of the Island is to the *Northward*, and bears from the anchoring Place of the said Port *E.* by *N.* and *W.* by *S.* Only take Heed here of the Shoal that is at the Windward End of *Callao*. The anchoring Place is any where right before the Houses. Tho' some Coasting-Pilots make this Island to lie *North* and *South*, it is *N. W.* and *S. E.* Here you may be furnish'd with whatsoever you stand in need of.

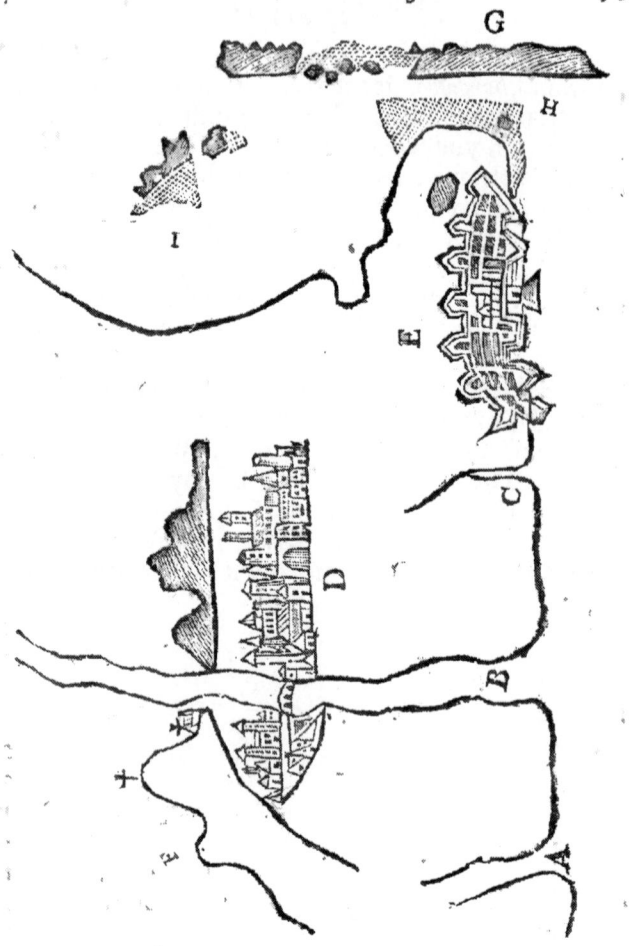

A. The River *Caravayllo*. B. The River of *Lima*. C. The watering Place. D. The City *Lima*. E. The Town of *Callao*. F. St. *Christopher*'s Hill. G. The Island of *Callao*. H. The Gut, or Paſſage. I. *Peña Oradada*, or, the pierc'd Rock.

A Description of Courses and Distances.

	Leagues.
From *Panama*, to Port *Perico*, S. W.	2
From *Panama*, to *Chepillo*, E. S. E.	7
From *Chepillo*, to the River *de la Maestra*, or Point Mangroves, S. E.	5
From the River *de la Maestra*, to that of *Chiman*, S.	4
From the River of *Chiman*, to that of *Mabe*, S.	2
From the River of *Mabe*, to *Pena Oradada*, or, the pierc'd Rock, S.	2
From *Pena Oradada*, to Point St. *Laurence*, S.	4
From Point St. *Laurence*, to Point *Garachine*, S. by W.	8
From Point *Garachine*, to the Island *de la Galera*, or, of the Galley, E. S. E.	5
From Point *Garachine*, to Port *Pinas*, S. by W.	7
From Port *Pinas*, to *Puerto Quemado*, or, burnt Haven, S. E.	12
From *Puerto Quemado*, to Cape *Corrientes*, or Currents, S. S. E.	20
From Cape *Corrientes*, to the Island *Palmas*, or, Palm-Trees, S. E.	10
From the Island *Palmas*, to the Bay of *Buena Ventura*, or, good Fortune, S. S. E.	10
From the Bay of *Buena Ventura*, to the Island *Gorgona*, S. W.	20
From the Island *Gorgona*, to the Island *del Gallo*, S. W.	24
From the Island *del Gallo*, to Point *Manglares*, or, Mangroves, S. S. W.	9
From Point *Manglares*, to the River of *Santiago*, S. W.	15
From the River of *Santiago*, to the Bay of St. *Matthew*, S. S. W.	5
From the Bay of St. *Matthew*, to that of *Tacames*, S. W.	3
From the Bay of St. *Matthew*, to Point *Galera*, or, of the Galley, S. W. by W.	6
From Point *Galera*, to Cape St. *Francis*, W. S. W.	2
From Cape St. *Francis*, to *Portete*, or, the little Port, S. E.	5

The South Sea Coasts.

Leagues.

From *Portete*, to *Barrancas Bermejas*, the red Crags, or, as others call it, *Vasia Borrachos*, S. W.   11

From *Barrancas Bermejas*, or, *Vasia Borrachos*, to Cape *Passado*, S. W.   6

From Cape *Passado*, to the Bay of *Caracas*, S. E.   4

From the Bay of *Caracas*, to the Port of *Manta*, S. W.   9

From the Port of *Manta*, to Cape St. *Laurence*, W. S. W.   4

From Cape St. *Laurence*, to the Island *Plata*, or, Plate, S. W.   4

From the Island *Plata*, to Point St. *Helena*, S.   18

From Cape St. *Laurence*, to Port *Callo*, S. E.   5

From Port *Callo*, to the Island *Salango*, S.   4

From the Island *Salango*, to the River *Colanche*, S. by E.   7

From the River of *Colanche*, to the Port of St. *Helena*, S. W.   3

From Port St. *Helena*, to the Point of the same Name, S. E.   1½

From Point St. *Helena*, to the River of *Chandui*, E. S. E.   8

From Point St. *Helena*, to the Island *Santa Clara*, S. E.   14

From the Island *Santa Clara*, to Point *Arena* in the Island *Puna*, E. N. E.   7

From the Island *Santa Clara*, to *Tumbes*, S. E.   4

From *Tumbes*, to Point *Mero*, S. W.   1

From Point *Mero*, to *Cabo Blanco*, or white Cape, S. W.   13

From *Tumbes*, to Cape *Blanco*, S. W.   14

From Cape *Blanco*, to Point *Parina*, S.   7

From Point *Parina*, to *Paita*, S. E.   7

From *Paita*, to *Pena Oradada*, or, the pierc'd Rock, S.   2

From *Pena Oradada*, to the Island *Lobos*, S.   2

From the Island *Lobos*, or, of Seals, to the Leeward Point *Aguja*, or, Needle, S.   12

From the Leeward Point *Aguja*, to the Windward Point *Aguja*, S.   4

From the Windward Point *Aguja*, to the Head-land of *Eten*, S.   19

Leagues.

From the Head-land of *Eten*, to the Hills of *Mozupe*, E. S. E. 4
From the Hills of *Mozupe*, to the Port of *Cheripe*, S. 3
From the Port of *Cheripe*, to *Pacafmayo*, S. E. 6
From *Pacafmayo*, to *Malabrigo*, S. E. 5
From *Malabrigo*, to *Guanchaco*, the Port of *Truxillo*, S. 11
From *Guanchaco*, to the Head-land of *Guanape*, S. S. E. 9
From the Head-land of *Guanape*, to that of *Chao*, S. E. 5
From the Head-land of *Chao*, to the Port of *Santa*, S. E. 4
From the Windward Point of the Island *Santa*, to *Fero*, S. E. by S. 1
From the Port of *Ferol*, to *Bonbacho*, S. E. by S. 4
From *Bonbacho*, to *Cafma*, S. E. 4
From *Cafma*, to *Mongon*, S. E. 3
From *Mongon*, to the Island of *Puerto Vermejo*, S. 4
From *Puerto Vermejo*, to the Port of *Guarmei*, S. E. 3
From *Guarmei*, to *Jaguei de la Zorra*, S. E. 4
From *Jaguei de la Zorra*, or, as others call it, *Haguei de la Cofta*, to the River *de la Barranca*, S. E. 9
From the River *de la Barranca*, to *Supe*, S. E. 2
From the Strand of *Supe*, to the Island of *Don Martin*, S. 2
From the Island of *Don Martin*, to the Port of *Guaura*, E. 1
From the Port of *Guaura*, to *Salinas*, or, the Salt-pits, S. E. 2
From the Head-land of *Salinas*, to *Maltefi*, S. W. 4
From the Rock of *Maltefi*, to *Tanbo de las Perdices*, E. 5
From the Rock of *Maltefi*, to *Hormigas*, S. 7
From the Head-land of *Salinas*, as above on the Continent, to *Tanbo*, or, *Playa de las Perdices*, S. 3
From *Playa de las Perdices*, to *Chancaillo*, S. E. 3
From *Chancaillo*, to *Chancai*, S. 2
From *Chancai*, to the Rocks *de los Pefcadores*, or, of Fishermen, S. 3
From the *Pefcadores* Rocks, to *Callao*, the Port of *Lima*, S. S. E. 5

CHAP.

CHAP. II.

The Sea-Coasts, &c. *from the Port of* Callao, *in the Kingdom of* Peru, *to those of* Caralmapo *and* Chiloe, *the most* Southern *in* Chile.

SHips sailing from *Callao*, the Port of *Lima*, above describ'd, to the Windward, must go out to the *Northward* of the Island of *Callao*; for they do not pass through the *Boqueron*, that is, the Mouth, Gut, or Channel, so call'd, which is between the Point of Land and the Island, because the Wind will not serve. But Ships coming from the Windward, to this Port, pass through the *Boqueron*, or Channel aforesaid; if they are small, they go in right before the Wind, and there is at least four Fathom Water in the shoalest Part of the Channel. The best Way is to stand in from somewhat to the Windward of the Island of *Callao*, and so pass on 'till the Part that is to the Landward, not that to Seaward, of a little Head-land there is without that call'd *Morro Solar*, be hid by the *Pena oradada*, or pierc'd Rock, which is a little Island you'll see there, about a League from the Continent, that is, all of it low there; and when the farthest inward Part, as I said above, of the little Head-land, where the Cross is, bears exactly with the *Pena oradada*, or pierc'd Rock, where the other Cross is, both Crosses being brought together, you may go in boldly without fearing any Thing, for provided you observe these Land-Marks, you are safe. At the End of the *Boqueron*, Channel, or Passage between the two great Islands, where the *North* Side of the greater Island begins, there also commences the shoalest Part of this Channel, which lies *N.W.* and *S.E.* and you must always keep your Poop towards *Pena oradada*, or the pierc'd Rock, as has

O 4 been

A Description of

been above directed; and when you perceive a great Rock, which lies between the two great Islands, is hid by the *South* Part of the Island, then keep closer to the Island, for then you are paſt the Shoal, and the Iſland has more Water than the Point of *Callao*, where the Bank lies, where you'll ſee the Sea-break. And when you are come up as far as a Break there is on the Iſland, with a white Spot at the Top of it, you may then incline a little towards *Callao*, ſtill giving a Berth to the Shoal; and come not very cloſe before the Point, when you fail in; and provided you keep the *North* Head-land of the Iſland to the *Weſtward*, you may ſafely go into the Port to anchor. Bring the two Croſſes together, as was ſaid above, you'll go ſafe, and then the Head-land call'd *Morro Solar*, and the Iſland *Pena oradada*, or, the pierc'd Rock, will ſhew as is here repreſented.

Morro Solar.

Pena oradada.

From the Head-land of the Windward Iſland of *Callao*, to the Port of *Paraca*, 35 Leagues *N.N.W.* and *S.S.E.* But to obſerve every noted Place in this Diſtance. From the aforeſaid Head-land of the Iſland, which is call'd la *Vieja*, or, the old Woman, to *Morro Solar*, two

The South Sea Coasts. 201

two Leagues. This is a high Head-land, which at a Distance in the Offing, shews a flat Table at the Top. From *Morro Solar*, to the Rocks of *Pachacama*, two Leagues. These are two great Rocks, and from the *South* Part of them runs a Ridge of little Rocks to the Continent, and they are all white.

Farellones, or Rocks of *Pachacama*.

From the Rocks of *Pachacama*, to the Point of *Chilca*, three Leagues. This is a low Point, with some bare Hillocks; and just at the Point rises a Ridge, running up the Inland, with three Breaks.

The Ridge of the Point of *Chilca*, up the Inland.

Whatsoever Way it be seen, it shews thus, bating that when you are near the Coast, either above or below, the Breaks appear narrower.

The

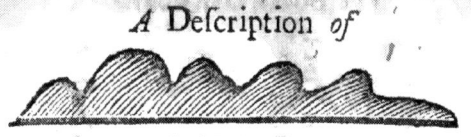

The same bearing *E. N. E.* shews thus.

From the Point of *Chilca*, to *Mala*, four Leagues; and from *Mala*, to the Island *Asia*, three Leagues. These seven Leagues make a Bay, and in the Midst of it three or four Hills, which, if the upper Land be cover'd with Fogs, looks at a Distance like black Islands, and are upright next the Sea; and the Shore from *Chilca*, to the Island *Asia*, lies *N. W.* and *S. E.* The Island *Asia* is white; and to the Landward of it, there are three Rocks, which are also white. This Island is in Compass about half a League; and as you come from above in the Offing, it has a Break in the Middle; and that Part of it which is to the *Southward*, is bigger and higher than the other Part to the *Northward*.

The Island *Asia*.

From the Island *Asia*, to *Canete*, or *Quenete*, or, as others call it, *Guarco*, seven Leagues *N. W.* and *S. E.* low Land next the Sea, and high a little up the Inland; and beyond that, is the *Cordillera*, or great Chain of Mountains. There is a large and deep Break, through which the River runs down, and forms on the Coast a little high Hill, which falls away towards the *S. E.* and ends over *Canete*. The Head-land of *Canete* is not very high, the Sea beats against it, and is frequently very high. On it are two Stone Forts built by the ancient *Indians*. You must anchor in nine Fathom Water.

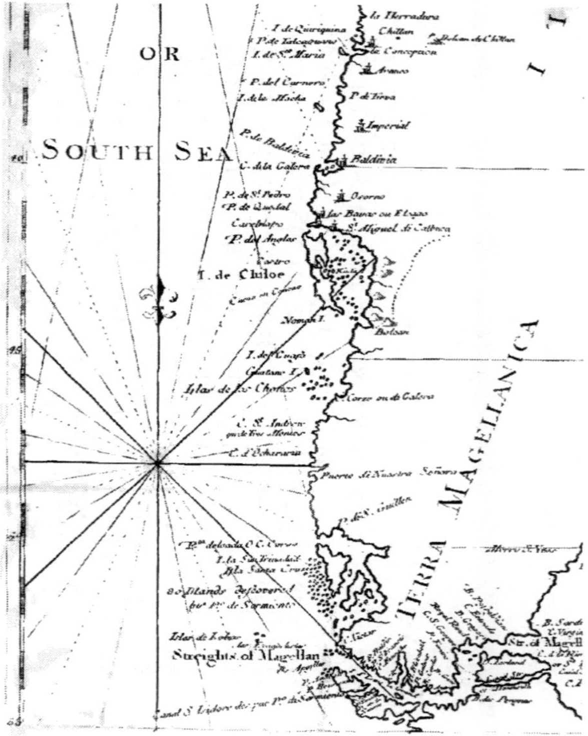

The Head-land of *Canete*,

Bearing N. N. E. shews thus.

The Break of *Canete*.

Coming from the Seaward, in somewhat above 13 Degrees of *South* Latitude, if the *Cordillera*, or great Ridge of Mountains, be clear, you will see this Break, which bearing E. N. E. at a great Distance, shews thus.

From *Canete*, to *Chincha*, nine Leagues, N. W. and S. E. low Land, with reddish and whitish Sloughs, and at the End of those Sloughs, to the S. E. is the Port of *Chincha*. If you would come to an Anchor there, you'll see a Palm-Tree a little up the Country, through a large *Guaca*, or eminent *Indian* Place of Worship, keep both of them right *East*; and when you are in seven or eight Fathom Water, you may drop your Anchor. Here the trading Vessels lade Corn and other Necessaries, some of them anchoring in six Fathom; but the King's Ships, which go to load Quick-silver at this Port of *Chincha*, to carry it to *Arica*, anchor farther out in nine Fathom, because there is generally a great Sea in the Port. Take heed therefore how you

you go afhore. The Town is inhabited by *Spaniards* and *Indians*, by whom you may be furnifh'd with all Neceffaries. Before this Port there are fix Iflands, always white, with the Dung of the Fowls that live on them; and about the Landing-Place there is Plenty of Fifh. They bear from the Anchoring-Place *N. E.* and *S. W.* diftant three Leagues, are low, and fomewhat reddifh, intermix'd with the white above-mention'd.

From the Port of *Chincha*, to that of *Pifco*, fix fmall Leagues *N. W.* and *S. E.* low Land, and in the mid Way is *Lorinchincha*, where the trading Veffels anchor, and load Corn and other Neceffaries. There is no Town, but all the Coaft is a fandy Strand, and there is a great Sea on it. The anchoring Ground is in fix Fathom Water, before a Houfe you'll fee there, and a white Church; which Place they call *el Molino*, that is, the Mill.

In the Port of *Pifco* the Ships anchor before the Houfes, in five or fix Fathom Water. In order to anchor here, you muft bring the Iflands of *Chincha* to bear *N. W.* the Ifland *de la Ballefta*, or, of the Crofs Row, *W. S. W.* and *Sangellan S. W.* There is faft clean anchoring Ground all along this Coaft backward, from *Canete* to *Pifco*, bating that about a League and a half fhort of *Pifco*, there is a white Hill or Ridge, call'd *Caucato*; come not too clofe to this Bit of the Coaft, becaufe of the River of *Pifco*, for there are fome Shoals running out from it. All along this Bay, which lies between the Iflands of *Chincha* and *Pifco*, being three Leagues in Length, Ships may ply upon a Wind, and anchor any where, for it is all clean. You may fail into this Port of *Pifco*, between the Iflands of *Chincha* and that of *Ballefta*, there being every where Water enough. At this Port there is Wood and Water, and all other Neceffaries. The Wind blows hard here from Noon forward.

Pifco,

Pisco, in 13 Deg. 40 Min. Latitude *South*.

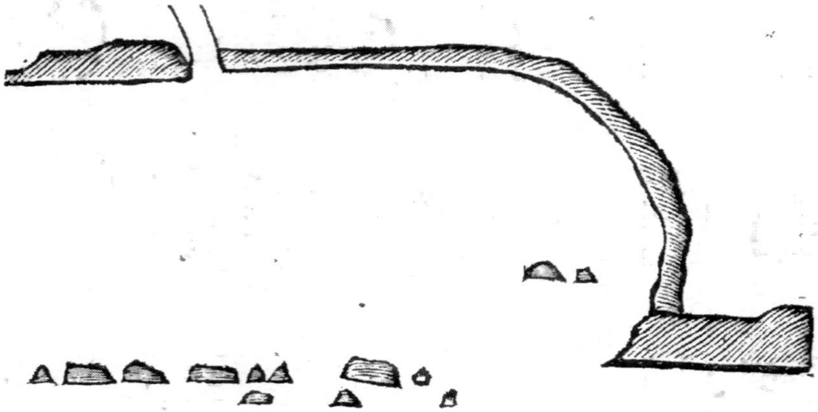

To the S. W. of *Pisco*, is the Island of *Sangallan*, high and thick, with a Break at the Top, and in some Positions, it shews two. The Continent opposite to this Island of *Sangallan*, is high, and call'd the Head-land of *Paraca*. All about the Island of *Sangallan*, is deep Water; on the *North* Side of it are two or three large Rocks, and some small ones, and some other small ones on the *South* Side. Between the said Island, and the Head-land of *Paraca*, is a deep Channel,

206 *A Description of*

Channel, through which the Ships pass which come from above. It is very deep, bating that close by *Sangallan* it has a little Shoal on the S. E. Side of it. This Island is in the Latitude of 14 Degrees *South*. It shews in several Shapes.

The Island *Sangallan*, in 14 Degrees Latitude *South*.

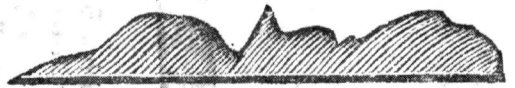

When you come from below, coasting along, and it bears from S. to S. E. distant six or eight Leagues, it shews thus.

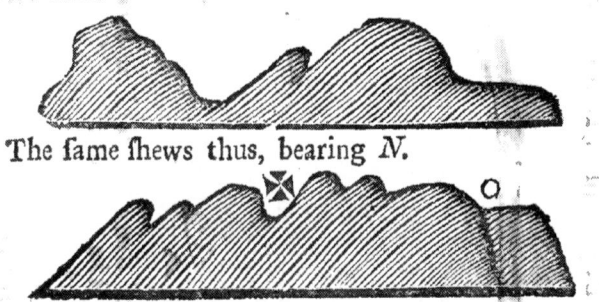

The same shews thus, bearing *N*.

This same Island of *Sangallan*, bearing *E*. shews thus, and as you leave it towards the *N. E.* the Head-land of *Paraca* will appear by Degrees, as represented at (o) and the Break at the Cross will close up. The Land of *Paraca* will appear to the *Eastward* of the Island.

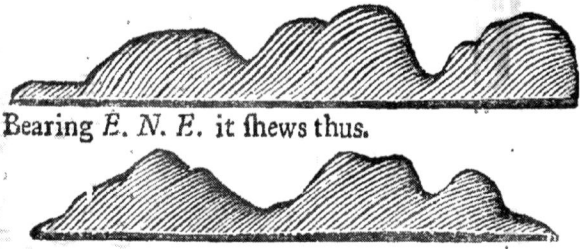

Bearing *E. N. E.* it shews thus.

When bearing *N. E.* at a great Distance, it shews thus.

Paraca,

The South Sea Coasts.

Paraca, above-mention'd.

When you come from the Seaward, in about 14 Deg. 30 Min. Latitude *South*, this Land of *Paraca*, bearing *N.E.* at a great Distance, will shew thus; and without it you'll see *Sangallan*, which shews a large Break when it bears *N.E.* very distant.

From *Pisco*, to *Paraca*, three Leagues; and this is the Port to *Chincha*, a *Spanish* Town, 15 Leagues up the Country. Here Ships use to careen, for it is a better Harbour than *Pisco*, and calmer; and here are the Store-houses of Wine which they bring from *Yca*. The anchoring is in five Fathom; and those Ships which cannot reach Port St. *John*, by Reason of the high Winds, put in here; for tho' there be Wind, there is no Sea; there is good anchoring Ground, and when once past the Rocks, Ships may ride any where. Those who fail out of this Port, are to observe, that there is a Shoal right before the Island *de la Ballesta*, towards the Head-land of *Sangallan*.

From *Sangallan*, to the little Island of *Sarate*, three Leagues *N. W.* and *S. E.* This little Island is round and low, the Continent opposite to it doubling, that is, one Land rising behind another, and full of Hillocks. The Island is about a quarter of a League from the Continent; and from it to *Morro de Viejas*, that is, the old Womens Head-land, is two Leagues *North* and *South*. This Head-land is high, and from the Seaward looks like an Island. The

South

208　*A* Description *of*

South Side is lower than the *North*, and on the Top of it is a Table, with a little Break, so hid, that it hardly appears; but in the Middle is a long and deep Break.

Morro de Viejas.

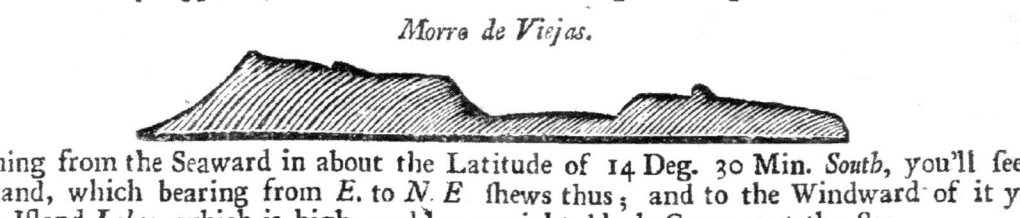

Coming from the Seaward in about the Latitude of 14 Deg. 30 Min. *South*, you'll see this Head-land, which bearing from *E.* to *N. E* shews thus; and to the Windward of it you'll see the Island *Lobos*, which is high, and has upright black Crags next the Sea.

The same Head-land to the Northward of the *N. E.* shews thus, at a great Distance, being high Land; but coming near, you'll see it rise and stretch out.

From this Head-land *Morro de Viejas*, to the Island *Lobos*, is little above half a League. The Island is high; there is anchoring on the *N. N. E.* Part of it, and to the *S. W.* it runs out in the Shape of a Galley, ending in a Sugar-loaf little Island, and beyond that another little Island, like a Ridge of Rocks, which all together at a Distance looks like Part of the Continent, lying close with the Sea and Crags.

At

At *Morro de Viejas*, or, old Womens Head-land, begins a great Bay, which ends at *Puerto Quemado*, or, burnt Head-land; and from the Island *Lobos*, or, of Seals, to *Morro Quemado*, is about half a League. *Morro Quemado*, or, burnt Head-land, is high and thick next the Sea, and within it the Land runs high and even, and generally this Head-land is cover'd with Fogs. If you would put into this Port of *Morro Quemado*, you must keep the Ship's Side close up with the Rocks, your Anchors clear, and your Fore-sail half loose; but if the Ship answers the Helm well, it is better to go in under a Sprit-sail, to be the clearer to moor, for there is much Wind, and great Squals; and if you can run up to the Port, which is where the Rocks end, and the Strand begins, keep close up with the Rocks, as is said above; and as soon as in the Port, drop your Anchor to the Landward, and moor Head and Stern, that the Ship may not wind. If you be not moor'd before the Wind rises, you'll be apt to drag, and always into deeper Water; therefore endeavour to get in betimes, before the Wind rises. When you would go out of this Port, you must pass between the Island *Lobos* and *Morro de Viejas*, which is all clean. In this Port there is neither Wood nor Water.

Morro Quemado, or, burnt Head-land, in 14 Degrees South Latitude, large,

Bearing *N.E.* shews thus.

From *Morro Quemado*, to *Punta de Olleros*, or Potters Point, six Leagues *N.W.* and *S.E.* the Land high and level. To the Leeward of this Point *Olleros*, there are some Rocks near the Continent, which look like little Islands. You may pass to the Leeward of them, to anchor in a little Bay, which is a good Harbour, in case of Squals, or strong Currents setting downwards.

This Port is not frequented, because there is nothing to lade.

Punta de Olleros, or, Potters Point.

Bearing *East,* distant five Leagues, shews thus.

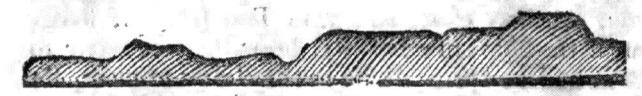

The same bearing *N. N. E.* shews thus; at the same Time with it will appear the Point of *Curacancana,* which we shall represent below.

From *Punta de Olleros,* or Potters Point, to Port *Caballa,* six Leagues *E. S. E.* and *W. N. W.* high Land next the Sea, with sandy Sloughs, and in the mid Way is a large Bay; and about a League and a half from Point *Olleros,* within the Bay, there are some upright Crags; and on the upper Part and Extremity of the level high Land, running from below, is a flat Table, call'd *Mesa de Dona Mariana,* that is, *Dona Mariana's* Table. From thence the Land falls as far as the River *Ica,* where it rises again high and plain, only at the River *Ica* there is a Piece of Land, forming a Table at the Top, with a Break on each Side; and through that to the Leeward, runs the River of *Ica.* Put not into this Bay, for there is a great Sea; and if the Wind falls, it will throw you on the Coast. The Port of *Caballa* has a lofty thick Head-land, which from the Offing shews plain at the Top. If you would anchor in this Port, you must always endeavour to make the Land to Windward of it, one Reason is, that you may keep clear of the Bay, and another, that you may furl your Sails in Time; and when you have so done, hoist out your Boat, and lower your

your Main Top-maſt, for the Wind is apt to blow hard, and there is a great Sea; therefore endeavour to come to an Anchor under a Sprit-ſail only. *Note,* That at the very Point of the Head-land, there is a ſharp-pointed Shoal, give it a Berth; then obſerve to keep a ſharp-pointed Rock, call'd *el Fraile,* or the Fryer, to the *Eaſt,* ſomewhat to Windward, inclining to *S.* by *E.* and when you ſee it bears with a little Head-land, which is above on the level, being the leaſt and moſt to Windward of three you'll ſee there; when the Fryer and this Hill are brought together, and you are in eight or nine Fathom Water, you may come to an Anchor. Obſerve, that, as I have ſaid, there is much Wind, and a great Sea in this Port, and the Port *de la Barca* is at the End of all the Rocks, where the Strand begins; and you muſt moor with two Anchors a-head, and a Kedger aſtern.

The Head-land of *Cavalla,* in bare 15 Degrees Latitude *South.*

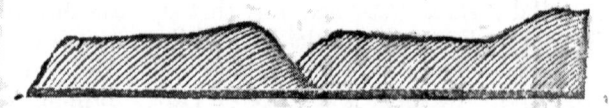

When the Break bears *N. E.* ſhews thus.

212 *A* Description *of*

Mesa de Doña Mariana, or, *Doña Mariana*'s Table.

When you come from above, or from the Seaward of Port *Cavalla*, the Land to Leeward *N. E.* from *Cavalla*, being the Table of *Doña Mariana*, will shew thus, at a good Distance, and you'll see this Coast lies almost *East* and *West*, and the Coast from above *N. W.* and *S. E.* and at the End of this Land is the Port.

From Port *Cavalla*, to Port St. *Nicholas*, five Leagues, higher Land than that of the Headland of *Cavalla*; and about a League and a half from the said Head-land of *Cavalla* to Windward, is a deep Break down to the Sea, through which the River of *la Nasca* runs. Farther to Windward in the higher Land, you'll see two white Sloughs, reaching from the Top to the Bottom, that which is to Windward the smaller of them. Over this high Land, as you come from the Offing, you'll see a Ridge of Mountains, Part whereof toward the *N. W.* is upright, or almost perpendicular, and toward the *S. E.* it grows slenderer, like a Galley, and has two or three small Breaks on the Top. This Ridge of Mountains is call'd *Curacangoma*, or *Curacancama*.

The

The South Sea Coasts.

The Mountains of Curacancama.

Coming from the Seaward, to make the Land about la Nasca, under 15 Degrees of South Latitude, when the Part at the Cross bears *East*, this Mountain of *Curacancama* shews thus.

If you come from the Seaward in upward of 15 Degrees from the Point of *Nasca*, when this Mountain of *Curacancama* bears N.E. it shews thus; and more to the *Northward* you'll see the Land of *Punta de Olleros*, or, Potters Point, and the Table of *Dona Mariana*.

The Port of St. *Nicholas* is very safe, but has neither Wood nor Water. Towards the S. E. of it is a low Point, which forms the Harbour; and over it is a topping round Hill, like a Sugar-loaf.

A Description of Point St. *Nicholas*,

Bearing *East*, shews thus.

From Port St. *Nicholas*, to Port St. *John*, two Leagues; but between these two Ports there are some red Sloughs. This Port St. *John* is a good Harbour, yet not resorted to, because there is no lading for Ships, nor has it Wood or Water. At the Entrance, to the Windward Side, is a Shoal, which is carefully to be avoided.

From Port St. *John*, to Port *Loma*, by others call'd *Acari*, six Leagues, low Land. The Head-land of *Cavalla* and Port *Acari* bear from one another N. W. and S. E. and the Head-land of *Acari* is about four Leagues up the Inland, being higher than that of *Cavalla*, having another Piece of Ground still higher. This Port of *Acari* is call'd the Port of *Loma*, or, of the Ridge, because it has a double Ridge of Land next the Sea, which forms the Port, and is a very good and clean anchoring Place, tho' not resorted to, because there is no Trade. All these three Ports of St. *Nicholas*, St. *John*, and *Loma*, or *Acari*, are very proper for Ships to take Shelter in, as they sail for *Arica*, or *Arequipa*, if they happen to meet with Squals, or Currents setting down. This Head-land of *Acari* and Port of *Loma*, are in 15 Deg. 30 Min. Latitude *South*.

The Head-land of *Acari*, Latitude 15 Deg. 30 Min.

Bearing *East*, distant four or five Leagues, shews thus.

The

The Land over Port *Acari*.

Coming from the Seaward in 15 Deg. 30 Min. Latitude, litttle over or under, you'll see this Head-land, and End of the high Land, which is over *Acari*; and when bearing N. E. it shews thus. Then the low Land runs about a League to the N. W. where it rises again the same Way on, somewhat double, looking at a Distance like an Island; and if it be clear Weather, you'll see the Head-land of *Acari*, which is generally under a Cloud or Fog.

From the Port of *Loma*, or *Acari*, to the Head-land of *Arequipa*, eight Leagues, low Land next the Sea, trending N. W. and S. E. This Head-land of *Arequipa* is high, and most of the Year under a Cloud or Fog. There is good anchoring, and the trading Boats usually lade there, nor is it so subject to high Winds as that of *la Nasca*.

The Head-land of *Arequipa*, in 16 Degrees, Latitude South.

Bearing from S. E. to S. shews thus.

The same bearing from N. W. to N. E. shews thus.
From the Head-land of *Arequipa*, to that of *Atico* 14 Leagues, N. W. and S. E. a little incilining to East and West. To the Windward of the Head-land of *Arequipa*, begins a sandy Shore, and runs on two Leagues to the Port of *Chola*, which is deep, being at the End of that Strand where a Parcel of large high Rocks rises. The Passage into this Port, is to the Leeward of those Rocks,

A Description of

Rocks, opposite to the End of the sandy Shore. If it be a small Ship, there is Shelter to the Leeward of those Rocks.

The Head-land of *Atico*,

Bearing *N.N.W.* shews thus.

Coming from the Seaward, in somewhat above 15 Degrees Latitude, you'll see this Land, being the Heights between *Arequipa* and *Atico*; and when the Heights at the Cross bear *E.S.E.* distant about ten Leagues, they shew thus; and if it be clear Weather, you'll see the Head-land of *Acari* to the *N.E.* and on the Coast the End of the high Land that comes to meet the low Land, running away to the *N.W.*

Lands between Atico and Arequipa.

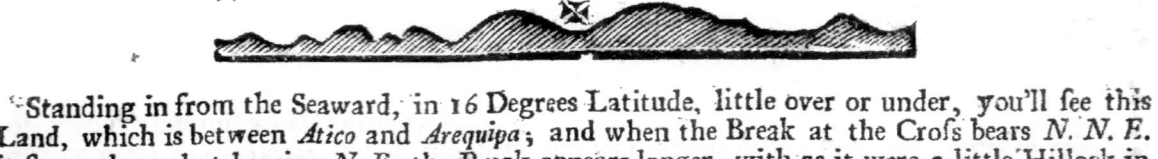

Standing in from the Seaward, in 16 Degrees Latitude, little over or under, you'll see this Land, which is between *Atico* and *Arequipa*; and when the Break at the Cross bears N.N.E. it shews thus; but bearing N.E. the Break appears longer, with as it were a little Hillock in the Middle.

From the Head-land of *Atico* to *Ocana* eight Leagues, N.W. and S.E. somewhat inclining to *East* and *West*, high Land next the Sea, and up the Inland snowy Mountains. Between *Atico* and *Ocana*, is a great Break made by a River running down to the Sea, and two Musquet-shot up the Break there is fresh Water. Near the Break there are two Rocks, call'd *Los Pescadores*, or, The Fisher-men.

From *Ocana*, to *Camana*, six Leagues, N.W. by W. and S.E. by E. a surly Coast, and *Camana* is a *Spanish* Town.

From *Camana*, to the Creek of *Quilca*, five Leagues, where small Ships put in; but great Ships come to an Anchor, when a Rock you'll see there bears *East*, half a League from the Creek. If you intend for this Creek of *Quilca*, and cannot get into the Harbour, by Reason of the Wind's falling calm, or the Current's setting down, there is good anchoring to the Leeward,

ward, in 20 Fathom Water. Let go your Anchor as foon as you fee the Strand of *Camana*, for it is every where clean and holding Ground. This Creek of *Quilca* is in 17 Degrees Latitude.

From the Creek of *Quilca*, to the Port of *Chule*, ten Leagues, *N. W.* and *S. E.* thus. From *Quilca*, to the Ifland *del Guano*, three Leagues. From the Ifland *del Guano*, to *Ilai*, four Leagues. Here they anchor within a Parcel of Rocks you'll fee, there is fo much Depth, that you muft be in above 40 Fathom Water. The Rocks are four or five, all white, and by them the Port of *Ilai* is known. There is no going into the Creek.

From *Ilai*, to *Chule*, which was once the chief Port to *Arequipa*, three Leagues, all the Coaft along nothing but Rocks; and here is a Creek, which Boats go into, being a meer Gut, and only one Boat can go in at once. When you would anchor, you muft open this narrow Creek, and you'll find 20 Fathom Water. If you are making from the Seaward for this Port of *Chule*, you'll fee the burning Mountain of *Arequipa* which bears with this Port *N. E.* and *S. W.* diftant 16 Leagues up the Inland; and if it be clear Weather, you'll fee other high Mountains, near that *Volcano*, which refembles a Sugar-loaf with the Top broken off.

The

The South Sea *Coasts.*

The burning Mountain of *Arequipa.*

Standing in from the Seaward, in 6 Deg. 30 Min. Latitude, for the Port of *Ilai,* or the Creek of *Quilca,* you'll see this Ridge of Hills, which, when the Clift at the Cross, being the burning Mountain of *Arequipa,* bears *N. E.* shews thus; and from the Ridge at (o) the Mountain runs away to the *N. W.* having near there another little burning Mountain. All these Hills and burning Mountains are cover'd with Snow, and have little Breaks.

Standing in from the Offing, in 17 Degrees Latitude, little over or under, if the *Cordillera,* or Ridge of Mountains, be clear of Clouds, you'll see these Hills, which are above *Arequipa*; and when they bear from *E.* to *N. E.* they shew thus; and the Clift at the Cross, is the burning Mountain of *Arequipa.* If you happen to be near the Coast, this burning Mountain shews in several Shapes.

From *Chule,* to the Port of *Ilo,* twelve Leagues *N. W.* by *N.* and *S. E.* by *S.* high Land, thus. Two Leagues to the Windward of *Chule,* is the River *Tanbo,* where is a Piece of low Land,

A Description of

Land, about a League in Length, all the rest steep and high. There is anchoring Ground before the River *Tanbo*, in 20 Fathom Water, a clean Bottom. From the River *Tanbo*, to *Terba Buena*, two Leagues. From *Terba Buena*, to the Port of *Ilo*, eight Leagues. If you would come to an Anchor in this Port of *Ilo*, you must make a Break in the highest Land, which you are to bring to bear *East*, as will the Vale thro' the Break, and then you may let go your Anchor. There is a good landing Place in this Port of *Ilo*, and close by it is a River of fresh Water. *Note*, That the Point of *Ilo* runs far out into the Sea, and is low; come not too near it, because of the In-draught. As you stand in from the Offing, you'll see this Point of *Ilo* low with the Sea, and at the Extremity of it is a little Island, which at a Distance seems to be four or five Rocks.

The Point of *Ilo*, in 18 Degrees *South* Latitude, large,

Bearing *N. E.* shews thus.
Point *Ilo*.

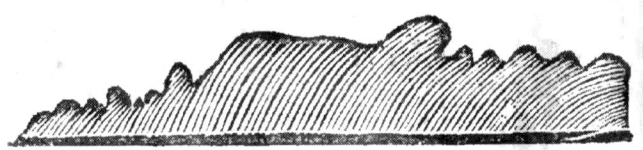

Coming from the Seaward, in 16 Degrees Latitude *South*, when near the Land, you'll see this Point of *Ilo*, which bearing *N. N. E.* will shew thus.

The same, when it bears *East*, shews thus; and then if you look out towards the *S. E.* you'll see the Headland of *Sama*.

From

The South Sea Coasts.

From the Point of *Ilo*, to the Head-land of *Sama*, eight Leagues *N. W.* and *S. E.* and by the Way is the Hill of *Acaguna*, and the River of *Ilo*, and the Hill call'd *Loma Quemada*, or, burnt Ridge. The River of *Ilo* is very good; and a Quarter of a League to the Windward, is the Town of *Ilo*, inhabited by *Indian* Fisher-men. Upon Occasion, you'll there find *Maiz*, or *Indian* Wheat, Water, and what else you want. There is a great Sea along this Coast, and in the River of *Ilo*. The Head-land of *Sama* is high and thick, and with the Coast makes several Appearances.

The Head-land of *Sama*, in 18 Deg. 30 Min. Latitude *South*, Bearing *East*, makes a Point, with several small Breaks, and shews thus.

The same bearing *N. N. E.* distant 10 Leagues, shews thus.

222   *A Description of*

The same, bearing *E. S. E.* shews thus.

The same bearing *N. E.* distant six Leagues, shews thus.

Loma Quemada, above-mention'd.

When you come from the Seaward, to make the Land to Windward of *Ilo*, you'll see this Land, call'd *Loma Quemada*, or, burnt Ridge; at the End it looks like the Mouth of a River, and then the high Land begins to rise again towards the *S. E.* and when the Ridge at the Cross bears *N. E.* it shews thus.

From the Head-land of *Sama*, to the Head-land of *Arica*, 13 Leagues *N. W.* and *S. E.* somewhat inclining *East* and *West*. By the Way is a large Bay, the greatest Part of it towards *Arica*, low Land, and a sandy Shore; only the Land of *Quiaca* is high. But in this Way; from the Head-land of *Sama*, to the River of *Sama*, three Leagues; and half a League to to Windward of this River of *Sama*, is the Port of *Quiaca*, high Land, where there are *Spaniards* and *Indian* Fisher-men; and there, upon Occasion, you may water, and be supply'd with what you want.

The high Land of *Sama*.

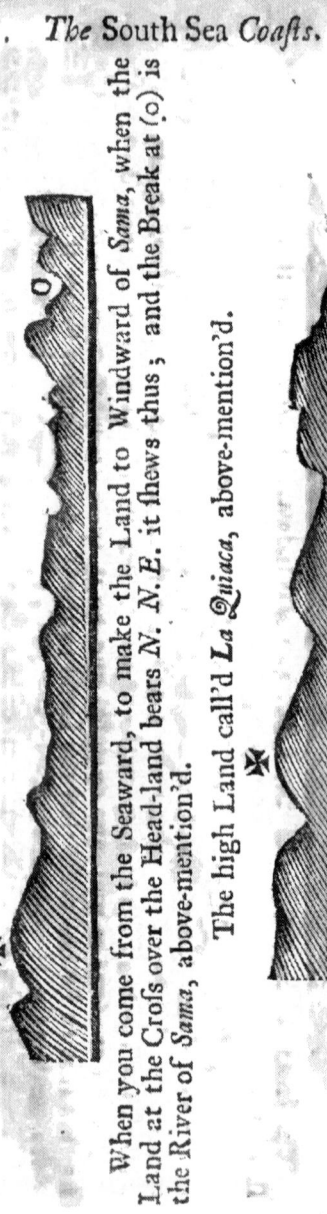

When you come from the Seaward, to make the Land to Windward of *Sama*, when the Land at the Crofs over the Head-land bears *N. N. E.* it shews thus; and the Break at (o) is the River of *Sama*, above-mention'd.

The high Land call'd *La Quiaca*, above-mention'd.

When the Height at the Crofs bears *East*, diftant feven Leagues, shews thus.

A Description of

The same, when the Height at the Cross bears *N. E.* distant seven Leagues, shews thus.

The same, when you make the Land to the Windward of *Arica*, and it bears *N. N. W.* shews thus.

From the Port of *Quiaca*, to the River of *Juan Diaz*, five Leagues; and from that River, to the Head-land of *Arica*, five Leagues, low Land, and a sandy Shore. Ships may anchor upon Occasion in this Bay, and along the Coast, for there is good holding Ground, and clean. *Note*, That the Sea runs high upon the Shore, if you happen to go in with your Boat. The Head-land of *Arica* is high and upright, and on it there are white Spots, being the Dung of Fowl. It is to be observ'd, that the Land to the Windward appears before this Head-land of *Arica*, as does the Head-land of *Sama*, and the Land of *Quiaca*, as being all higher than it, and because it lies up at the Bottom of the Bay. When they came to an Anchor in this Bay formerly, they us'd to open the Street call'd *de la Merced*, but now they anchor more to Windward. As soon as past the little Island, you may bend towards the Land; and when you have discover'd all the Ware-houses, and the first Street in the Town, which is call'd
del

del Tanbo on the Shore begins to open, and you are in eight or nine Fathom Water, you may let go your Anchor, keeping the little Iſland a-head. Here you muſt moor with the Kedger a-ſtern, becauſe of the Land-Breezes, which blow hard in this Bay. This Port of *Arica* is in bare 19 Degrees Latitude *South*. If you be coming in from the Seaward, and have not yet had Sight of the Coaſt, by Reaſon of the Diſtance, or any other Cauſe, if the Weather be clear, and the Mountain-Land appears, you'll ſee on the Top of the *Cordillera*, or Ridge of Mountains, two Hills, which look like Rocks in the Sea, or burning Mountains, with the Snow upon them. If theſe two Hills bear *N. E.* from you, then you are ſomewhat to Windward of the Port ; and to the *S. E.* of the ſaid two Hills, on the ſame Ridge of Mountains, is another Hill, looking white like the others with the Snow.

The Head-Land of *Arica*, in 19 Degrees, Latitude *South*.

When you come coaſting along from the Windward of *Arica* downwards, this Head-land bearing *N. N. E.* ſhews thus.

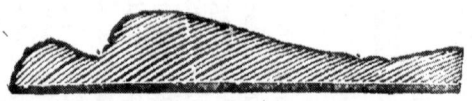

The ſame bearing *N. E.* ſhews thus.

From the Head-land of *Arica*, to that of *Tarapaca*, 5 Leagues, high Land next the Sea, lying *North* and *South*, a little inclining *Eaſt* and *Weſt*, and it is known by its trending, becauſe the Land to Leeward of *Arica* lies *N. W.* and *S. E.* Between *Arica* and *Tarapaca*, there are three deep Breaks made by Rivers running down to the Sea. If when you come in from the Seaward, for want of an Obſervation, or by Reaſon of the Cur-

Vol. II.	Q	rents,

226 *A Description of*

rents setting down, you happen to make the Land of *Arequipa*, take Heed you are not deceiv'd by some other Breaks there are on that Coast, tho' you may discern the Difference, by Reason that Land is lower, and lies *N.W.* and *S.E.* whereas this to Windward of *Arica* lies *North* and *South*. The first Break to Windward of *Arica*, is that of *Vitor*, which others call of *Corpa*, and it is five Leagues from *Arica*. About a League to the *Northward* of this Break, almost at the Edge of the Water, there are white and red Crags, for about a League in Length, and look like a Wall, because that Part of the Coast is perpendicular. To the *Southward* of this Break, is a Head-land, with some Crags that are white from Top to Bottom.

o   The Break of *Vitor*.   ☩

When you make the Land, somewhat to Windward of *Arica*, and the Head-land at (o) which is to Windward of the Break of *Vitor*, bears *East*, somewhat *Northerly*, and the Head-land at the Cross is to the Leeward of the Break of *Camarones*, to be mention'd below.

The same Land, distant better than four Leagues, when the Break at the Cross bears E. N. E. shews thus.

From the Break of *Vitor*, to that of *Camarones*, seven Leagues; and to the *Southward* of this Break of *Camarones*, close by it, is a little Rock, very white; which, when you are about five

Leagues

Leagues from the Shore, looks like a Ship under Sail. You must always be near this Coast to know it; and when the two Breaks above-mention'd bear *East*, little over or under, they will be open to you, which is not so with the other more to Windward, call'd of *Pisagua*, which forms a Sort of Bay, the *North* Point whereof winds to Windward, and the Channel must bear N. E. to have it open to you.

The Break of *Camarones*, in 19 Deg. 30 Min. Latitude *South*.

When you have made the Coast to Windward of *Arica*, which runs almost *North* and *South*, the Break of *Camarones* at the Crofs bearing E. N. E. about eight Leagues distant, shews thus with the Head-land at (o).

The burning Mountains of *Tacoral*.

If you happen to stand in from the Seaward in about 49 Degrees Latitude, and the *Cordillera*, or, great Ridge of Mountains, is clear, so that you can make it, and not the Coast, when the burning Mountains at the Crofs bear N. E. they shew thus; and you are to Windward of *Arica*.

From

228 A Description of

From the Break of *Camarones*, to that of *Pisagua*, eight Leagues, high Land next the Sea, and running *North* and *South*, like the rest.

Pisagua.

When you come from the Seaward, in about 19 Degrees and a half Latitude, little over or under, you'll see this Land; and at first Sight the Head-land at (o) will look like an Island, but drawing nearer, you'll see all the Land and Coast to Windward of *Arica*, which lies between the Break of *Camarones*, and that of *Pisagua*, and at the Cross is *Pisagua*.

Pisagua.

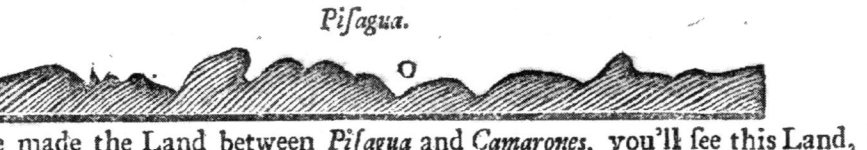

When you have made the Land between *Pisagua* and *Camarones*, you'll see this Land, being the Break of *Pisagua* at the (o) bearing E. S. E. distant 17 Leagues, when it shews thus, and the Break of *Camarones* will be open to you, and to the Windward of it a white Spot, with two or three little black Streaks in the Middle of it, from the Top to the Bottom.

To

The South Sea Coasts. 229

To the Windward of Pisagua.

When you come from the Seaward, to make the Land to Windward of *Pisagua*, you'll see this Land, which bearing *E. N. E.* shews thus. The Head-land at the Cross is to Windward of the Mouth of the Break, and close by the said Head-land you'll see some Rocks, and to the Windward of it a little white Spot; and the Land farther up above the Break, is higher than that next the Sea, and here and there you'll see Hills of the *Cordillera*, or great Chain of Mountains, and among them the Hill at (o) and sometimes it will appear over the Break, and other Times to Windward with another on the Top of the Chain of Mountains, and they are both cover'd with Snow.

From the Break of *Pisagua*, to the Head-land of *Tarapaca*, six Leagues. This Head-land is high, and from it runs low Land every Way, the Head appearing in whatsoever Position you are to it. The Coast is boisterous, and on it is a little Island.

The

230 A Description of

The Head-land of *Tarapaca*, in 20 Deg. 40 Min. Latitude *Sout*.

When you make the Land between *Pisagua* and *Tarapaca*, and the Head-land of *Tarapaca* bears S. E. it shews thus at a Distance; and tho' the Point at (o) looks low, near at Hand you'll find it as high as the rest of the Coast.

When you make the Land even with the Height of the Head-land of *Tarapaca*, and it bears near about *East*, it shews thus; and if you look towards the S. E. you'll see the Head-land of *Pica*, which makes two slender Skirts, or Descents, still falling away, the one to the *Northward*, and the other to the *Southward*, and the Head-land on the Top has a little Break, not deep.

The same, when you come to make the Land between *Pica* and *Tarapaca*, and this Head-land bears E. N. E. being somewhat to Leeward, shews thus; and to the Windward is the Head-land of *Pica*, with the two Skirts aforesaid.

From

The South Sea Coasts.

From the Head-land of *Tarapaca*, to that of *Pica*, five Leagues *North* and *South*. Near it is a little white Island, and to the Landward of it there is good anchoring in 7 Fathom Water.

The Head-land of *Pica*,

Bearing *E. S. E.* shews thus.

When you come from the Seaward, in about 20 Deg. 30 Min. Latitude, little over or under, at a great Distance this Head-land of *Pica* will shew near thus, howsoever you be posited.

But when you draw nearer this Head-land of *Pica*, and it bears *East*, it will shew thus, with these little Breaks, and the Coast runs away high, both to the *Northward* and to the *Southward*.

From

From the Head-land of *Pica*, to the River of *Lora*, 12 Leagues, high Land, lying *North* and *South*, a bold Coast, with white Crags. The River lies where the Land is lowest and drawn in, which must bear *East* if you would make it. The Water of this River is somewhat brackish. The anchoring Ground is a quarter of a League to Windward, near a Parcel of small Rocks, which are to be a-head of you.

From the River of *Lora*, to *Atacama*, 15 Leagues, high Land, lying *N. N. E.* and *S. S. W.* the Coast very rocky and deep; but keep clear of it. Five Leagues from *Lora*, *Southward*, and before a Ridge of Mountains, is *Haguey de Paquisa*, where there is fresh Water. If you would water here at *Paquisa*, you must anchor before a Head-land, which forms as it were a Point, with some white Spots, and then you'll see a great Tree before the *Jaguey*, or watering Place; and if right, you'll be before the highest Land of this Coast; and over the Point is a Hill, which, when you are out at Sea, looks like three or four Hills, and is full of Thistles.

This Hill of *Paquisa*, being in 22 Degrees Latitude *South*,

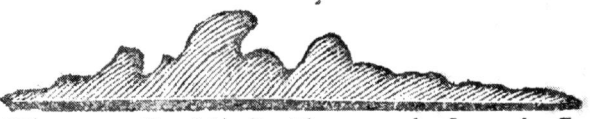

When you stand in directly upon it from the Seaward, shews thus.

In case you miss of Water at *Paquisa*, it is but two Leagues from thence to *Algodonales*, where there is fresh Water, to be known by some white Spots next the Sea.

From *Algodonales*, to *Atacama*, eight Leagues, where Water may also be had, but it is somewhat brackish.

From *Atacama*, to the Bay of *Mijillones*, or Muscles, five Leagues, *N. E.* by *N.* and *S. W.* by *S.* At the Point of it is a high round Hill, like a Sugar-loaf, and on the *North* Side of it another smaller. The Bay of *Mijillones* is deep, the anchoring Place is to the *Eastward*,

The South Sea *Coasts.*

ward, and the Entrance lies *North* and *South.* Here is no Water. Ships may also anchor to the *South* of the Point, near a Rock that is to be seen there, in 15 Fathom Water.

The Hill forming the Point at the Bay of *Mijillones*,

When it bears from *East* to *S. E.* shews thus.

From the Point of the Bay of *Mijillones*, to *Morro Moreno*, or brown Head-land, eight Leagues, *N. E.* and *S. W.* This Head-land is high and thick, and there is anchoring on the *North* Side of it, near a little Island, where Ships may be shelter'd, when there are Squals from the *South*, and be in a Readiness to sail. Here is a good Port, tho' streight, and Ships may careen. There is Water, but it can only be taken up at low Ebb, because it has a Communication with the Sea upon the Flood. If you coast along, the Land will appear to you stretch'd out and plain next the Sea, like a little Table over the Point. When you come from above, *Morro Moreno* will appear high and round. Here is generally very blowing Weather in this Latitude of 23 Degrees.

Morro Moreno, or brown Head-land,

Bearing from *South*, to *S. W.* shews thus.

From *Morro Moreno*, to *Morro de Jorge*, or *George's* Head-land, 10 Leagues, *N.* by *E.* and *S.* by *W.* and between them is form'd a Bay, which is dangerous when there is much Wind at *S. W.* because that blows right in, and it is unsafe to be in it at that Time. There is anchoring at the Point of *Morro de Jorge*, in

25 Fathom, good holding Ground. This Head-land and *Sangallan*, lie *N.* by *W.* and *S.* by *E.*

From *Morro de Jorge*, to *Baia de Nueſtra Senora*, or our Lady's Bay, 20 Leagues, *N. N. E.* and *S. S. W.* very high deſert Mountains, without any Water, bating that about ſix Leagues ſhort of our Lady's Bay, there is ſome freſh Water; but it is in the higheſt Part of the Mountain, and in the very midſt of it towards the Bay. Below the Place of the Water, there is a plain Spot down next the Sea, you may come to an Anchor before that Plain; for it is all clean, and there is much Water cloſe under the Shore, ſo that you muſt anchor in 25 Fathom. Note, that a Ridge runs down from the Mountain to the Sea; and where the Ridge terminates, is a white Rock about half a League from the Continent. This Rock muſt be to the *Southward* and a-head of you, and you are to be a Mile from it. This white Rock is in 24 Degrees and a half Latitude *South*. From this ſame Rock, at the watering Place, you may ſee *Morro Moreno*, which thence looks high and round.

From this white Rock at the watering Place, to the *Baia de Nueſtra Senora*, or our Lady's Bay, one half of the Country inhabited, and the other half deſert; and here ends the Mountain, and begins the low Land. Here alſo the boiſterous *North* Winds begin to reign. This Bay of our Lady is very deep, and the Coaſt ſo ſteep, that you'll ſcarce ſtrike Ground with 50 Fathom Line, a very blowing Coaſt when there are Squals.

From *Baia de Nueſtra Senora*, or, our Lady's Bay, to the Head-land of *Copiapo*, thirty Leagues *N. N. E.* and *S. S. W.* From our Lady's Bay, to Port *Betas*, 6 Leagues, good anchoring Ground, deep Water, and you muſt ride in 30 Fathom, to be ready to ſail when the *North* Wind blows. You'll know this Port of *Betas* by a Spot of white Sand, and in the midſt of the White another black Spot, and at the upper Part of the Port there are ſome Chaps or Clefts, which look like the Veins in a Mine, call'd in *Spaniſh*,

Betas,

Betas, whence the Name of the Place. Here is no fresh Water. The Latitude is 26 Degrees *South*.

From Port *Betas*, to *Juncal*, or, the rushy Ground, six Leagues. This is but a bad Harbour; for it is only shelter'd against the S. W. with Mountains, the Country desert, and no fresh Water.

From *Juncal*, to Port *Cheveral*, six Leagues, a good Port, and has an Island without, which secures it against all Winds; but it has no fresh Water.

From *Cheveral*, to the Head-land of *Copiapo*, 12 Leagues, and, by the Way, good Places to anchor, and Bays shelter'd from the *South*; but a desert Country, and without fresh Water. The Port of *Copiapo* is good, secure against the *South*, as also against the *North* Winds. The Head-land of *Copiapo* somewhat resembles an Island, is not unlike Point *Santa Elena*, and has a little Island on the *South* Side, about a League from the Head-land. There is also good anchoring on the Side of this Island next the Continent, near a high Head-land that is there to be seen. Ships may sail from this Port with a *North* Wind. The Country here is inhabited.

The Head-land of *Copiapo*,

Coming from the *Northward*, shews thus.

From the Head-land of *Copiapo*, to the Islands *de Pajaros*, or of Birds, 33 Leagues, N. E. by N. and S. W. by S. thus. From the Island of *Copiapo* to *Baia Salada*, or the Salt Bay, five Leagues. Here is good Anchoring and Water, but that none of the best. The River of *Copiapo* is between the Island and the Bay; and to the Seaward of this River, is a dangerous Shoal, bearing *West* from the River.

From *Baia Salada*, or Salt Bay, to *Totoral*, 10 Leagues, where the anchoring Ground is to the *Northward* of the Point, which is to bear from you S. W. by W. You must hold your self ready to sail, if the

North

North Wind should rise, which blows in. Here you may water.

From *Totoral*, to the Port of *Guasco*, 10 Leagues; and here is Shelter from the *East* to the *S. E.* This is a Vale well peopled, with a River. You must anchor opposite to this River, near a little low Island. There are seven or eight Rocks at the Windward Point opposite to the Port, and on the very Point is a sandy Hill, with a Break at the Top, and Sands about on both Sides. There are three little Islands at the anchoring Place. This Hill of *Guasco* is high, thick, and round, and makes a Saddle at the Top, and it is higher to the *Northward* than to the *Southward*, running off low on both Sides.

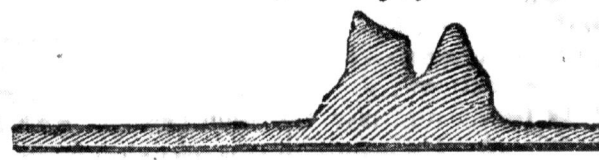

The Hill of *Guasco*,

From the Seaward, shews thus.

From this Port of *Guasco*, to the Islands *de Pajaros*, or of Birds, eight Leagues. To the Landward of the greater Island, is a little one somewhat low, where you may anchor, and sail out again with a *North* Wind which Way you please. There are four great Islands, and the middlemost, which is the largest, has five Breaks on it. That which lies most to the Landward, has some high Rocks about it; and the two great Islands that lie together, bear from one another *N. N. W.* and *S. S. E.*

From these Islands *de Pajaros*, or of Birds, to the Port of *Coquinbo*, seven Leagues, *N. N. W.* and *S. S. E.* This Port of *Coquinbo* has a Point which is not very high, and at the Entrance into it are two Rocks opposite to the Point. These Rocks, if you would go in, you must leave on your Starboard Side, and 'till they

bear

bear S.S.W. you'll find no Ground; for which Reason, you must make up close to the Point, and when near it, stand in to anchor before the highest Land in a Line, with a Rock there is in the Sea, call'd *la Tortuga*, that is, the Tortoise. From this anchoring Place, to the Sea-Port Town of *Coquinbo*, is two Leagues, and it is in 30 Degrees of *South* Latitude.

From the Point of *Coquinbo*, to the Bay of *Longoi*, seven Leagues, N. E. and S. W. and one League to Windward of *Coquinbo*, is a Port call'd *la Herradura*, that is, the Horse-shoe, which is safe and good anchoring Ground.

From the Bay of *Longoi*, to *Puerto del Governador*, or the Governor's Port, 24 Leagues, N. N. E. and S. S. W. thus. From *Longoi*, to *Limari*, eight Leagues. This Port of *Limari* is known by the Woods of very tall Trees on the Mountains, and in the midst of them a deep Break down to the Sea.

From *Limari*, to *Choapa*, 10 Leagues. The Land at *Choapa* high Mountains, most Part of the Year cover'd with Snow; the Coast upright, and without any Port.

From *Choapa*, to *Puerto del Governador*, or, the Governor's Port, six Leagues. This Governor's Port is good; you must anchor opposite to the lowest Land, and hold your self ready to sail when the *North* Wind comes up, right before a high Hill that has a small Break on the Top, and on the Side of the Hill towards the S. W. is a Tuft of Trees. The Port has a little Island; and, in case you cannot weather the Point with the *North* Wind, you may take Shelter to the Leeward of the little Island. This Governor's Port is in 31 Deg. 15 Min. Latitude *South*.

From the Governor's Port, to that of *Valparaiso*, 20 Leagues *North* and *South*. This Port and Point *Coroma*, bear from one another N. by E. and S. by W. This same Port and *la Ligua*, bear from one another N. W. and S. E. distant five Leagues. At this Port of *Ligua*,

near

near the Point, is a Shoal two Fathom under Water, and there is five Fathom Water at the anchoring Place.

From *la Ligua*, to *Puerto del Papudo*, three Leagues, the anchoring Place deep, and at it is a high Hill, with a Break on the Top, and a Spot of Trees, which looks like the Hill at the Governor's Port.

From Port *Papudo*, to the Flats of *Quintero*, five Leagues. These Flats are above Water, near the Continent, and the Sea sets very much towards them. Ships may safely pass between these Flats and the Continent, where there is twelve Fathom Water, clean Ground.

From these Flats, to Port *Quintero*, two Leagues. This Port *Quintero* is deep, and safe against the *Southerly* Winds, but the *North* blows directly in at the Mouth; take heed you be not catch'd there by it.

From Port *Quintero*, to the Port of *Valparaiso*, five Leagues, *N. E.* and *S. W.* By the Way, three Leagues from *Quintero*, is the River of *Chile*, North and South; and at the River of *Chile* is a Shoal, on which you'll immediately see the Sea break.

From the River of *Chile*, to *Valparaiso*, the Port to the City of *Santiago*, Capital of the Kingdom of *Chile*, two Leagues; and in this small Space between the aforesaid Places, there are three several Strands, and down the middlemost runs the River *de las Minas*, or, of the Mines, which others call of *Margamarga*; and at the End of the last Strand, to the Windward, is this Port of *Santiago*, call'd *Valparaiso*, where there is a Break and a little open Strand, and at the End of that Strand is a stony Point. You must come to an Anchor behind that Point, before the little Strand, which makes the Shelter from the *North* Wind, under the highest Land, which runs towards the Point of *Coroma*. There is seven Fathom Water in the anchoring Ground; when moor'd, one of your Anchors must be almost ashore, and the other to the *Northward* near a Parcel of Stones

on

on the little Strand near the Shore, good holding Ground. However, you muſt keep your ſelf in a Readineſs to ſail, becauſe the *South* Wind blows hard, and there is a rowling Sea, eſpecially after Noon. This Port of *Valparaiſo* is in 32 Degrees *South* Latitude, large.

From the Port of *Valparaiſo*, to the Point of *Coroma*, two Leagues. The Courſe *W. S. W.* On the Inſide of this Point, is a Shoal; come not too near in your Way from above for *Valparaiſo*. There is anchoring at this Point of *Coroma*. This Point and the Iſlands of *Coquinbo*, the Head-land of *Copiapo* and *Morro Moreno*, or brown Head-land, all bear from one another alike, that is, *N. E.* and *S. W.*

From the Point of *Coroma*, to *Potocalma*, 18 Leagues *N. E.* and *S. W.* and ſix Leagues to Windward of the Point of *Coroma*, are *las Salinas*, or, the Salt-Pits, where there is anchoring Ground, with a *South* Wind, the Land low, with many Shoals, and you muſt anchor in 10 Fathom.

You may anchor within the Head-land of *Potocalma* with a *South* Wind; and a very little League within it the Wind blows hard, and the Sea is boiſterous, and ſo deep, that very near the Land you'll ride in 25 Fathom Water, clean Ground. This Head-land of *Potocalma* is in bare 34 Degrees Latitude *South*.

From the Head-land of *Potocalma*, to the Break or River of *Lora*, 14 Leagues. Here at *Lora* is a large Break, like that of the Iſland of *Lima*, and a Piece of the Coaſt is low, being ſandy, lying *N. E.* and *S. W.* and right before the Middle of the higher ſandy Ground, is good anchoring. At *Lora* the Mountains begin to have Trees on their Tops; ſo that there is Plenty of Wood, and ſo all the Coaſt continues to the City of the *Conception*.

From the Break or River of *Lora*, to the River of *Maule*, ſeven Leagues. This River has three Fathom Water at low Ebb, on the Bar; and there are two

Rocks

Rocks at the Entrance; and about half a League to the Leeward of them, there is anchoring, against the *South* Wind. This River of *Maule* is so boisterous, as if the *North* Wind were produc'd in it.

From the River *Maule*, to Point *Humos*, 10 Leagues. This is a dangerous Point, for from it several Shoals run out, and some Ships have been lost at it. The River *Maule*, and this Point, bear from one another *N. N. E.* and *S. S. W.*

From Point *Humos*, to the River of *Itata*, seven Leagues. The Country is very populous about this River of *Itata*, and there is anchoring at a Point which runs out, and it makes a deep, upright, and very large Break. From Point *Humos*, to *la Herradura*, or, the Horse-shoe, there is no Bottom, unless it be in these Parts here mention'd.

From the River of *Itata*, to *la Herradura*, or, the Horse-shoe, five Leagues. This Port is a Bay, in Form like a Horse-shoe, whence it has the Name. At the Entrance there are three or four high Rocks, and within it is land-lock'd against both the *North* and *South* Winds.

From *la Herradura*, or, the Horse-shoe, to the Island of the *Conception*, commonly call'd *la Quiriquina*, two Leagues *N. E.* and *S. W.* These two Leagues make the Passage, or Mouth for going up to the *Conception*, and this Passage lies *North* and *South*; so that the *North* Wind blows right in. The *Spanish* Town lies along the Sea, which there forms a spacious sandy Shore. A quarter of a League from the City of the *Conception*, is a River, call'd *Andalin*, where the trading Boats can run up. In this Port of the *Conception*, you must anchor before a small Rivulet, that runs through the Middle of the City, and ride out at some Distance from it, to be ready to sail, in case the *North* Wind should come up, when you are to make for the Point of *Talcaguano*, and there come to an Anchor to Leeward of a low Point, so that *Talcagnano* Point and the Island *Quiriquina*

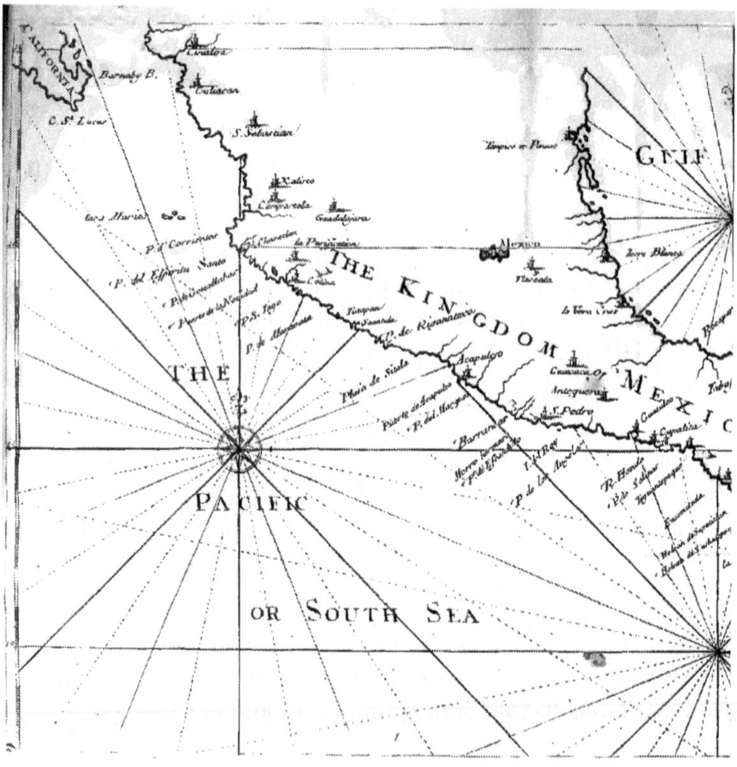
CALIFORNIA
Barnaby B.
C. St. Lucas
Cinaloa
Culiacan
S. Sebastian
Xalisco
Compostela
Guadalajara
Tampico or Panuco
GULF
THE KINGDOM OF MEXICO
Mexico
Tlascala
la Vera Cruz
Acapulco
Antequera
S. Pedro
THE
PACIFIC
OR SOUTH SEA

The South Sea *Coasts.*

riquina may bear equal from each other; and then you'll be shelter'd from the *North* Wind. At this Island *Quiriquina* there is clean and safe anchoring on both Sides of a sharp Point you'll see there. The Island and the City of the *Conception* bear from one another *W.* by *N.* and *E.* by *S.* distant two Leagues, and the Port is in 36 Deg. 15 Min. Latitude *South.* The Ships that are bound for *Valdivia* and *Chiloe* from the *Conception,* go to anchor at Point *Talcaguano,* in 12 Fathom Water, to wait for the *North* Wind, and must leave the Rock call'd *Farellon de las Ollas* about half a League to the *East,* as also the Channel which runs between Point *Talcaguano,* and the Island *Quiriquina,* through which Ships do not pass up, by Reason of its Narrowness, unless they have the Wind right astern.

The Island *Quiriquina,*

When you are standing in from the Seaward, shews thus.

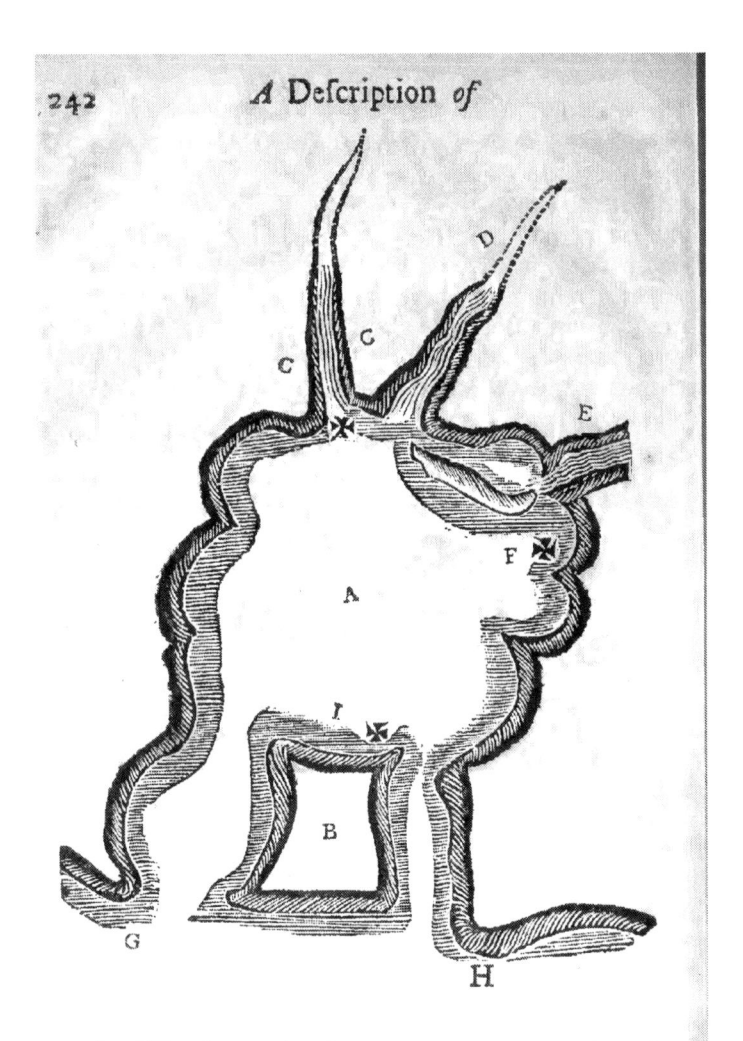

A. The Bay of the *Conception*. B. The Island *Quiriquina*. C. The Town of the *Conception*. D. The River *Andalita*. E. *Itata* River. F. Port *Talcaguano*. G. Point *Herradura*. H. Point *Talcaguano*. I. Port *Quiriquina*.

From

From the Point of *Talcaguano*, to St. *Mary*'s Ifland, 11 Leagues *N. E.* and *S. W.* thus. From Point *Talcaguano*, to Port St. *Vincent*, two Leagues; and this is a good Port, fafe againft all Winds, except the *Weft*, which blows right in.

From Port St. *Vincent*, to the River of *Biobio*, two Leagues. This is a mighty River, at the Mouth of it are two high Rocks; and between it and Port St. *Vincent*, is a high Hill, which looks like two Dugs, and is therefore call'd *Las Tetas de Biobio*, that is, the Dugs of *Biobio*.

Tetas de Biobio,

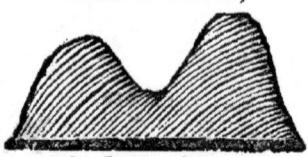

From the Seaward, fhews thus.

From the River *Biobio*, to *Lavapie*, feven Leagues; and this is a Bay affording Shelter againft the *South* Wind, but the *North* blows right in. Here is the Ifland of St. *Mary*, where there is anchoring Ground on the Side next the Continent, and Shelter againft the *North* Wind, but little Water. Near this Ifland is a high Rock, and the *North* Point is low, with anchoring Ground about it, but not clear of Rocks; and on the Outfide is a Shoal running about half a League out to Sea; take Heed of it, for fome Ships have perifh'd on it. This Shoal and the Dugs of *Biobio* bear from one another *N. E.* and *S. W.* The Dugs are high, and may be feen at a great Diftance, making a large Break in the Middle, as reprefented above. The Ifland of St. *Mary* is inhabited, plain Land, and about two Leagues in Compafs.

From St. *Mary*'s Ifland, to the Ifland *Mocha*, 20 Leagues *N. N. E.* and *S. S. W.* thus. From St. *Mary*'s Ifland, to *Puerto del Carnero*, or, the Sheep's Port, 10 Leagues. Near this Port is a high Rock, and in it

R 2 there

there is fresh Water and a River; and hither the Boats resort with Provisions for the Garrisons.

From Port *Carnero*, to the Island *Mocha*, 10 Leagues; but from Port *Carnero*, to *Tucapel*, four Leagues. This is no good Port, and worst of all when the *North* Winds prevail along the Coast. The Island *Mocha* is high Land, well peopled by *Indians*, come over from the Continent, which is four Leagues distant, and to the *W. S. W.* there are Shoals.

The Island *Mocha*, in 38 Deg. 40 Min. Latitude *South*,

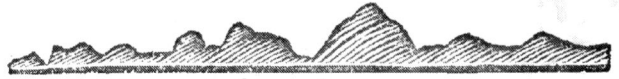

To the Leeward, shews thus.

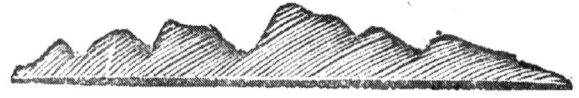

The same to the Windward, shews thus.

This Island and the Port of *Valdivia* bear from one another *North* and *South*, distant 25 Leagues. The Island and the River *Imperial N. W.* by *W.* and *S. E.* by *E.* the same and Point *Galera N.* by *W.* and *S.* by *E.* the same and the Island *Juan Fernandes N. W.* by *W.* and *S. E.* by *E.* distant 90 Leagues.

From the River of *Tucapel*, to the River *Imperial*, 10 Leagues. This River has good Depth of Water, the Land is low, and you may anchor in it, for it is clean.

From the River *Imperial*, to *Alquivite*, 10 Leagues, low Land; and between these two Places, is a Hill, which falls away both to the *Northward* and the *Southward*, down to the very low Coast; for such it is all the Way to near *Valdivia*.

The Hill between *Imperial* and *Alquivite*,

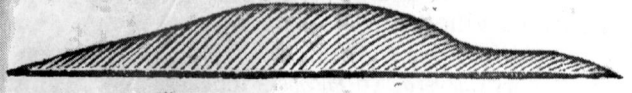

Shews thus in any Position whatsoever.

Note, That the lowest Land on this Coast, is at *Alquivite*; and if it be clear Weather, you'll see some burning Mountains up the Country, and then the Land rises to the *Southward*, as far as the Head-land of *Bonifacio*, which is 10 Leagues.

This Head-land of *Bonifacio* is the one Side of the Entrance into the Port of *Valdivia*, the other is *Morro Gonzalo*, or, *Gonzalo*'s Head-land, opposite to it on the other Side, and three Leagues distant; and this Head-land of *Gonzalo* is perpendicular, but not very high, and has some white Spots upon the very Point, or End of it.

The Head-land *Morro de Bonifacio*,

When bearing *N. N. E.* shews thus.

246 *A* Description *of*

The Head-land *Morro Gonzalo,*

When bearing *E. N. E.* shews thus.

Between these two Head-lands, as I have said, is the Passage up to *Baldivia*; and from the Head-land of *Bonifacio*, to the Port *del Coral*, is five Leagues *N.* by *W.* and *S.* by *E.* and from the Head-land of *Gonzalo*, to the Port *del Coral*, is two Leagues, and somewhat better. The Course between this same Mouth of the Harbour, and the Point of *Niebla*, is *N. W.* and *S. E.* and when you are up as far as the Point of *Niebla*, in the Midst of the River, you may bend to the *Southward*, and you'll presently see the Bay and Port of *Coral*, and may come to an Anchor where you please, near the Shore, in six or seven Fathom, for it is all clean; and at the End of the Strand towards the *S. E.* is the watering Place, tho' there is also Water in several other Places on the same Strand to the *Southward*. Take Heed how you go there, for there are unconquer'd *Indians*, and they are very treacherous. You'll find them all about that Side, and must not trust them if you go ashore, but stand always upon your Guard, whilst the Men fill the Water, and look out sharp. Take Heed your Boat be not aground, left a great Number of *Indians* attack you, that you may not want that Retreat in case of Need, and let your Oars be clear. I am particular in this Point, lest you come by the

the same Misfortune, as has befallen others, of being surpriz'd. Here has been a Fort built some Years since, which has render'd this Place the safer, most of the *Indians* about having submitted.

If you would proceed from the Port *del Coral*, to the Island *Constantina*, which you will see towards the *E. S. E.* you may proceed boldly, for all the Way there is five Fathom Water, half a Fathom over or under; but be sure both in entering this River, and running up it every where, to sound all the Way, and then come to an Anchor on the *East* Side of *Constantine's* Island, which will be on your Starboard-Side. When within the Island, you'll anchor in four Fathom, or four and a half, in which Position the *King's* Island will bear from you *East*, somewhat *Northerly*. This is a large Island, and the Channel lying between it and the Continent to the *Northward*, is that the Boats pass up to *Valdivia*, being three Leagues. There is but little Water in this Channel, not above three Fathom at most, and in some Places only one and a half. By the *South* Channel there are five Leagues from the Mouth of it to the Town; and this Way Ships go up, because there is four or five Fathom Water. On the *S. E.* side of the Island, is a Bay of little Depth; go not into it with a Ship. The Draught here represents all the Parts of this Harbour, from the Mouth upwards, and the Depth of Water in every Part.

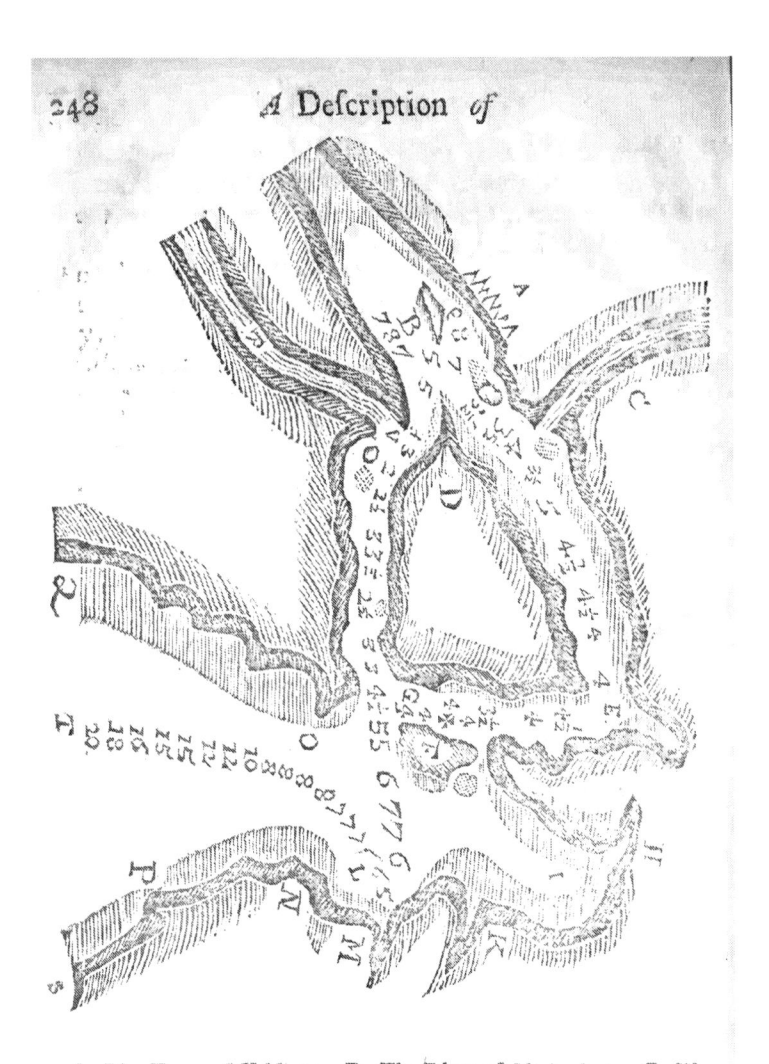

A. The Town of *Valdivia*. B. The River of *Mariquina*. C. The River *Callacalla*. D. The King's Island. E. The Passage or Channel for great Ships. F. *Constantine*'s Island. G. The Passage or Channel for Boats. H. The River *Claro*. I. St. *John*'s Bay. K. S. *Christopher*'s River. L. Port *Corral*. M. Watering Place. N. Port *Amargos*. O. Point *Niebla*. P. The Head-land *Morro Gonzalo*. Q. The Head-land *Morro Bonifacio*. R. *Valdivia* River. S. Point *Galera*. T. The Mouth of the River. V. *Mota* Island.

From

The South Sea *Coasts.* 249

From Port *Coral*, to the Head-land *Morro Gonzalo*, two Leagues, and from *Morro Gonzalo*, to the Point *de la Galera*, or of the Galley, three Leagues, E. by N. and W. by S. the Land doubling, or appearing one above another, only Point *Galera* is low next the Sea, and then the Land rises a little to the *Southward*.

Point *Galera*,

Bearing S. S. W. shews thus.

The same bearing N. N. E. shews thus.

From Point *Galera*, to *Rio Bueno*, or good River, five Leagues, high Land next the Sea, and the River makes a large Break above.

From *Rio Bueno*, or, Good River, to *Puerto de san Pedro*, or St. *Peter*'s Port, nine Leagues, high Land like the last; and at this Port, is another Break like that of *Rio Bueno*.

From St. *Peter*'s River, to the Point of *Quedar*, eight Leagues, the same Sort of Land as above. Between Point *Galera* and the Point of *Quedar*, the Coast runs N. N. E. and S. S. W. 22 Leagues. This Point of *Quedar* appears in several shapes, according to your Position.

The Point of *Quedar*,

Bearing S. by E. shews thus.

The same bearing S. S. E. shews thus.

The

The same bearing *S. E.* shews thus.

From the Point of *Quedar*, to Point *Godoi*, four Leagues. There are three Rocks bearing with this Point *N. E.* and *S. W.* Ships may safely pass close by them, for there is no Danger.

From Point *Godoi*, to the *Bahias*, or, Bays of *Lago*, four Leagues *North* and *South*. None should put in here, without Necessity obliges; and if any do go in, let them keep up close to the *North* Side, and not to the *South*, because there are Sands which run out very far.

From these Bays, to the Port and Fort of *Caralmapo*, three Leagues. Here the Vessels put in with Necessaries for the Garrisons. To go safe into this Harbour, they must cling close under the Shore, because else the Current, which sets out between the Continent and the great Island of *Chiloe*, and is very strong in that Channel, will be apt to drive them out. This Port of *Caralmapo*, is a small Bay, in 42 Deg. 30 Min. Latitude *South*, and in it you must anchor very close to the Shore.

From this Port of *Caralmapo*, to the nearest Land of the great Island of *Chiloe*, two Leagues, *North* and *South*. This Island is very large, being above 22 Leagues in Length, lying *North* and *South*. At the *North* End is a Point call'd *del Anco*, and at the *South* End another call'd of *Quela*. All the Side of the Island towards the Sea, is Crags, except in the Middle, where is the Port of *Cucao*, over which are two high Hills, much alike, resembling Dugs, and therefore call'd *las Tetas de Cucao*, that is, the Dugs of *Cucao*. The Port is small, and therefore not resorted to. At the Point *del Anco*, there is a very good Harbour, Landlock'd against all Winds, call'd *Puerto del Ingles*, or the *English* Man's Harbour; but none use it, because foul.

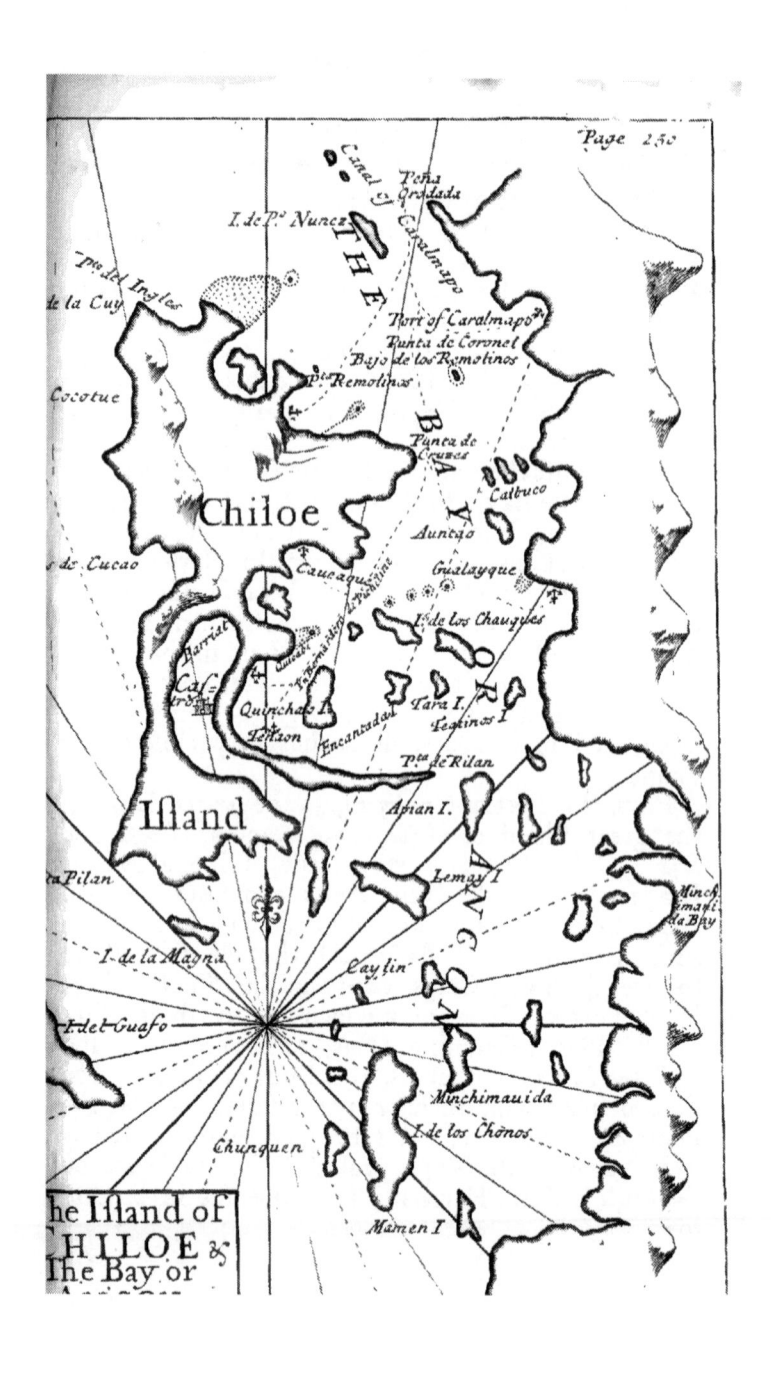

The Island of
CHILOE &
The Bay or

At the *South* End of this great Island, are three smaller, the largest of them call'd of the *Magdalen*, and there the Bay runs in and forms a spacious Coast, and in it is a great Number of Islands. On the *East* Side of it, is the *Spanish* Town of *Castro de Chiloe*, where the Ships from *Peru* load Timber, whereof there is great Plenty, and several Ships are built here. If you would go in, it must be at the *South* Channel, between the great Island and the Island of *Guafo*, which Channel is 10 Leagues wide, the Distance between the two Islands, which is a safe Channel, and you may ply upon a Wind in it.

The Island of *Guafo* is four Leagues in Length, having two Points, the one to the *N. E.* and the other to the *S. E.* for so the Island is posited. Take heed of this *S E.* Point, for there are Shoals running out from it half a League to Sea, on which some Ships have perish'd. The middle Part of the Island of *Chiloe* is in 43 Degrees, Latitude *South*.

From the Island *del Guafo*, to *Cabo de Tres Montes*, or, Cape Three Mountains, where the great Bay, form'd by the Island of *Chiloe*, terminates, is 10 Leagues; so that the whole Extent of the said Bay is 96 Leagues. Cape *Tres Montes*, and the Land of it, lyes *N. E.* and *S. W.* is high and mountainous, and at the End of the Cape it forms three Points, and therefore it has the Name of Cape Three Mountains or Hills.

From Cape *Tres Montes*, to Cape *Corzo*, 63 Leagues, *North* and *South*, in which Distance there are some Bays and Rivers, but never a Rock, nor Island.

At Cape *Corzo*, the Land turns away to the *East* 36 Leagues, forming a Nook call'd *Ancon sin salida*, that is, without a Thorough-fare. On the *North* Side of it, are high Mountains and two spacious sandy Bays, and at the End of the Mountain there are two small Rivers, and at the very Extremity of this Nook a mighty River falls into the Sea, where the Coast turns again to the *S. W.* 62 Leagues, that is, to the

Mouth

Mouth of the Streights of *Magellan*. In this Space there are some Bays, and three Rivers, and in this Bay there are eleven Islands, two large, and nine small. The great one to the *Northward*, is call'd St. *Mary*'s Island, and about it are the four smallest of the others. The other great Island, which lies nearer the *Magellan* Streights, is St. *Martin*'s Island, and about it are five other small Islands. There is a Passage to the Landward of all these Islands.

At the very Mouth of the Streights of *Magellan*, to the *Northward*, are four Rocks together, call'd the Four Evangelists. *Magellan*'s Streight is in 52 Degrees of *South* Latitude.

To the *Southward* of the Streights, close to the Mouth, are other twelve Rocks call'd the twelve Apostles, and from this Place the Coast winds away *Eastward* 125 Leagues, the Distance from the Streights of *Magellan*, to those of *le Maire*, or, as the *Spaniards* call them, of St. *Vincent*, the Breadth whereof is eight Leagues large, and the Length of the Passage five. Streight *le Maire* is in 55 Degrees, Latitude *South*.

Courses and Distances.

	Leagues.
From the Head-land of the Windward Island of *Callao*, call'd *la Viejas*, or, the old Women, to *Morro Solar*, S. S. E.	2
From the Head-land *Morro Solar*, to the Rocks of *Pachacama*, S. S. E.	2
From the Rocks of *Pachacama*, to the Point of *Chilca*, S. S. E.	3
From the Point of *Chilca*, to *Mala*, S. S. E.	4
From *Mala*, to the Island *Asia*, S.	3
From the Island *Asia*, to *Canete*, S. E.	7
From *Canete*, to *Chincha*, S. E.	9
From the Port of *Chincha*, to that of *Pisco*, S. E.	6
From *Pisco*, to *Paraca*, S. E.	3

The South Sea *Coasts.* 253
                                                    Leagues.
From *Sangallan,* the Head-land at Port *Paraca,* to the
   Island *Sarate, S. E.*   3
From the Island *Sarate,* to *Morro de Viejas,* or, Old Wo-
   mens Head-land, *S.*   2
From *Morro de Viejas,* to the Island *Lobos, S.*   ½
From the Island *Lobos,* to *Morro Quemado,* or, Burnt
   Head-land, *S.*   ½
From *Morro Quemado,* to *Punta de Olleros,* or, Potters
   Point, *S. E.*   6
From *Punta de Olleros,* to Port *Caballa, E. S. E.*   6
From Port *Caballa,* to Port St. *Nicholas, S. E.*   5
From Port St. *Nicholas,* to Port St. *John, S. E.*   2
From Port St. *John,* to Port *Loma,* or *Acari, S. E.*   6
From Port *Loma,* or *Acari,* to the Head-land of *Are-
   quipa, S. E.*   8
From the Head-land of *Arequipa,* to that of *Atico, S. E.*
                                                         14
From the Head-land of *Atico,* to *Ocana, S. E.*   8
From *Ocana,* to *Camana, S. E.* by *E.*   6
From *Camana,* to *Quilca, S. E.*   5
From *Quilca,* to the Island *del Guano, S.*   3
From the Island *del Guano,* to *Ilai, S. E.*   4
From *Ilai,* to *Chule, S. E.*   3
From *Chule,* to the Port of *Ilo, S. E.* by *S.*   12
From the Point of *Ilo,* to the Head-land of *Sama, S. E.*
                                                          8
From the Head-land of *Sama,* to the River of *Sama, S.
   E.*   3
From the River of *Sama,* to the Port of *Quiaca, S. E.* ¼
From the Port of *Quiaca,* to the River of *Juan Diaz,
   S. E.*   5
From the River of *Juan Diaz,* to the Head-land of *Ari-
   ca, S. E.*   5
From the Head-land of *Arica,* to the Break of *Vitor* or
   Corpa, S.   5
From the Break of *Vitor,* to that of *Camarones, S.*   7
From the Break of *Camarones,* to that of *Pisagua, S.*   8
                                                       From

Leagues.
From the Break of *Pisagua*, to the Head-land of *Tarapaca*, S. 6
From the Head-land of *Tarapaca*, to that of *Pica*, S. 5
From the Head-land of *Pica*, to the River *Lora*, S. 12
From the River *Lora*, to *Atacama*, S. S. W. 15
From *Atacama*, to the Bay of *Mijillones*, S. by W. 5
From the Bay of *Mijillones*, or Muscles, to *Morro Moreno*, S. W. 8
From *Morro Moreno*, or, Brown Head-land, to *Morro de Jorge*, S. by W. 10
From *Morro de Jorge*, or, *George*'s Head-land, to *Baia de Nuestra Senora*, S. S. W. 20
From *Baia de Nuestra Senora*, to Port *Betas*, S. S. W. 6
From Port *Betas*, to *Juncal*, or, Rushy Ground, S. S. W. 6
From *Juncal*, to Port *Cheveral*, S. S. W. 6
From *Cheveral*, to the Head-land of *Copiapo*, S. S. W. 12
From the Head-land of *Copiapo*, to *Baia Salada*, S. W. by S. 5
From *Baia Salada*, or, Salt Bay, to *Totoral*, S. W. by S. 10
From *Totoral*, to the Port of *Guasco*, S. W. by S. 10
From the Port of *Guasco*, to the Islands *de Pajaros*, or, of Birds, S. W. by S. 8
From the Islands *de Pajaros*, to the Port of *Coquimbo*, S. S. E. 7
From the Point of *Coquimbo*, to the Bay of *Longoi*, S. W. 7
From the Bay of *Longoi*, to *Limari*, S. S. W. 8
From *Limari* to *Choapa*, S. S. W. 10
From *Choapa*, to *Puerto del Governador*, or, the Governor's Port, S. S. W. 6
From the Governor's Port to *la Ligua*, S. E. 5
From *la Ligua*, to *Puerto del Papudo*, S. E. 3
From Port *Papudo*, to the Flats of *Quintero*, S. E. 5
From those Flats, to Port *Quintero*, S. E. 2
From Port *Quintero*, to the River of *Chile*, S. E. 3

The South Sea Coasts.

Leagues.

From the River of *Chile*, to the Port of *Valparaiso*, S. E.   2

From the Ports of *Valparaiso*, to the Point of *Coraoma*, W. S. W.   2

From the Point of *Coraoma*, to the Head-land of *Potocalma*, S. W.   18

From the Head-land of *Potocalma*, to the River of *Lora*, S. by W.   14

From the Liver of *Lora*, to the River of *Maule*, S. by W.   7

From the River of *Maule*, to Point *Humos*, S. S. W.   10

From Point *Humos*, to the River *Itata*, S. S. W.   7

From the River *Itata*, to la *Herradura*, or, the Horse-shooe, S. S. W.   5

From la *Herradura*, to the Island of the *Conception*, commonly call'd la *Quiriquina*, S.   2

From Point *Talcaguano*, opposite to the Island *Quiriquina*, to St. Mary's Island, S. W.   11

From Point *Talcaguano*, to Port St. *Vincent*, S.   2

From Port St. *Vincent*, to the River *Biobio*, S.   2

From the River *Biobio*, to *Lavapie*, S. by W.   7

From St. *Mary*'s Island, to the Island *Mocha*, S. S. W.   20

From St. *Mary*'s Island, to *Puerto del Carnero*, or, Sheeps Port, S. E.   10

From Port *Carnero*, to the Island *Mocha*, S. W.   10

From Port *Carnero*, to *Tucapel*, S. by W.   4

From *Tucapel*, to the River Imperial, S.   10

From the River Imperial to *Alquivite*, S.   10

From *Alquivite*, to the Head-land of *Bonifacio*, S. by W.   10

From the Head-land of *Bonifacio*, being the *North* Side of the Entrance into the Port of *Valdivia*, to *Morro Gonzalo*, or, *Gonzalo*'s Head-land, which is at the *South* Side, S. by E.   3

From the Head-land of *Bonifacio*, to the Port *del Corral*, S. by E.   5

From

                                                         Leagues.
From the Head-land of *Gonzalo* to the same Port *del Coral*, E.   2
From the Head-land of *Gonzalo*, to Point *Galera*, or, of the Galley, W. by S.   3
From Point *Galera*, to *Rio Bueno*, or, Good River, S.   5
From *Rio Bueno*, to *Puerto de san Pedro*, or, St. *Peter's* Port, S.   9
From St. *Peter's* River, to Point *Quedar*, S. S. W.   5
From Point *Quedar*, to Point *Godoi*, S. W.   4
From Point *Godoi*, to the Bays of *Lago*, S.   4
From the Bays of *Lago*, to the Port of *Caralmapo*, S.   3
From the Port of *Caralmapo*, to the nearest Land of the great Island *Chiloe*, S.   2
From the *North*, to the *South* Point of the Island *Chiloe*,   22
From the *South* Point of *Chiloe*, to the Island *Guafo*, S.   10
From the Island *Guafo*, to Cape *Tres Montes*, or, Three Mountains, S. E.   10
From Cape *Tres Montes*, to Cape *Corzo*, S.   63
From Cape *Corzo*, to the Nook call'd *Ancon sin Salida*, S. E.   36
From *Ancon*, to the Cape, at the Mouth of *Magellan's* Streights, S. W.   62

CHAP.

CHAP. III.

The Sea-Coasts, &c. from the Port of Panama, *on the* Isthmus *of* America, *to that of* Acapulco, *in the Kingdom of* New Spain, *and thence to* California.

FROM *Panama,* to Port *Perico,* two Leagues, as has been said above, and they bear from one another, *N. E.* and *S. W.* and in the Mid-way is a dangerous Shoal, which bears *North* and *South* with *Paitilla.*

From Port *Perico,* to *Otoque,* four Leagues, coasting along the Shore at a convenient Distance.

Otoquillo.   *Otoque.*

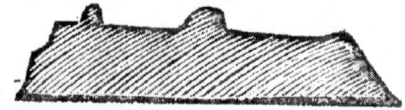

These two Islands of *Otoque* and *Otoquillo,* that is, little *Otoque,* bearing *S. W.* shew thus.

When you are come up the Length of *Otoque,* steer away *S. S. W.* for *Morro de Puercos,* or, the Head-land of Swine, and the Point *de Hignera,* or of the Fig-tree; and when you are up with *Otoque,* you'll see the Headland of *Chame,* where a spacious Bay runs in, being that of *Nata.* Venture not in, lest the Wind should blow up it, which may endanger your Ship, and no getting out. This Bay ends at the Island *Iguanas,* to the Leeward of which there is good Anchoring, and Shelter from the *S. W.* Wind, and to the Windward of it is also anchoring Ground at the Continent, in a Creek, the Shore whereof is all a white Sand, and here also is good Shelter against the *S. W.* Wind. From *Otoque,* to this Island *Iguanas,* is two Leagues.

A Description of
The Island *Iguanas*,

Bearing S. W. shews thus.

From the Island *Iguanas*, to *Punta Mala*, two Leagues S. W. To the Leeward of this Point, is anchoring Ground in a Cove the Continent makes, safe against the S. W. Wind, clean Bottom, and ousy. At this Point, the Land trending *North* and *South* with *Nata* terminates.

From Point *Mala*, to Point *Higuera*, seven Leagues, N. W. This Point in a Ridge, which runs tapering out to the Sea S. E. to the Leeward of it is anchoring Ground, and Shelter against the S. W. Wind. If you would get up to this anchoring Place, you must keep close under the S. W. Head-land, and may ride where you think fit; for it is all clean Bottom, and you'll find Wood and Water. Note, that two Leagues short of Point *Higuera*, there are two small Islands call'd *los Frayles*, or, the Fryers, little above half a League distant from each other. The nearest to the Continent, is about a League distant; the other more to the Seaward, is higher, round, and bare at the Top; and without this bare Rock, is a Sand under Water, on which the Sea usually breaks with a S. W. Wind.

Point *Higuera*, and the Islands call'd the Fryers,

Bearing W. by N. shew thus.

From Point *Higuera*, to *Morro de Puercos*, or, the Head-land of Swine, two Leagues, W. by N. Between the said Point and Head-land, is a Sand under Water, on which the Sea breaks with the S. W. Wind. It is

about

The South Sea Coasts. 259

about a League from the Continent, and all round the Point is flat, and you'll find 15 Fathom Water a League from the Shore, all Sand. If you want to take Shelter against the *S. W.* Wind, there is anchoring Ground to Leeward of *Morro de Puercos.* You'll there fee a Bay it forms, before a fandy Shore. The anchoring Place is near a Rock there to be feen, which Rock is to be kept a-head. At the Entrance into this Place, there is very deep Water, and Ships ride within in 15 Fathom.

Morro de Puercos, or, the Head-land of *Swine,*

Bearing *W. S. W.* diftant feven Leagues, fhews thus, and the Mountains of *Gunete* rife on to the *Weftward.*

The fame bearing *W. by S.* fhews thus.

The fame bearing *North,* fhews thus.

If you would ſtrike over from *Morro de Puercos*, to the Coaſt of *Peru*, at the Time when the Trade-Winds reign, you muſt ſteer from this Head-land, to *Malpelo*, S. S. E. and from *Malpelo* South, to the Coaſt of *Peru*. This Iſland of *Malpelo* is little and high, and when bearing *Eaſt*, ſhews many little Breaks; bearing S. W. ſhews a Break in the Middle, not very deep; and from the *North*, to the N. E. it ſhews round. If it be the Seaſon when the S. W. Winds reign, and you are forc'd to ply upon a Wind, make the moſt of the Trip to the *Weſt*, rather than that to the S. E. becauſe the Coaſt you are deſign'd for lies N. E. and S. W. and the more you gain towards the S. W. the more you'll be to the Windward of *Malpelo*. If the ſtrong Currents ſhould happen to carry you to the Iſland *Gorgona*, you may know it by two Breaks it makes, with a Peek in the Middle, being all high Land, the Windward Head-land larger than the other to Leeward, and you'll ſee no other Land about it.

The Iſland of *Malpelo*,

Bearing S. E. diſtant three Leagues, ſhews thus.

The ſame bearing S. W. diſtant four Leagues, ſhews thus.

The ſame bearing *South*, diſtant five Leagues, ſhews thus.

The South Sea Coasts.

From *Morro de Puercos*, the Mountains of *Guanico*, go rising to the *Westward*, 'till they terminate at Point *Mariato*. From *Morro de Puercos*, to this Point, is 12 Leagues; the Coast lying *East* and *West*, is steep and rocky. The Mountains are high, and full of Heads or Hillocks, and on the highest Part is a large round Hill, and a Break on the *West* Side. About these Mountains there are frequently fierce S. W. Winds, mighty Squals, and a high Sea.

The Mountains of *Guanico*,

Bearing *N. N. E.* distant 12 Leagues, shews thus.

The Point *Mariato* is high, mountainous, and wooded, and has a little Rock to the *Westward*, close by it, which is also full of Trees. The Coast is upright, deep, and rocky.

Point *Mariato*,

Bearing *N. N. W.* distant eight Leagues, shews thus, with the low Island without the Rock.

If

If you are bound from *Morro de Puercos,* for *Nicaragua,* your Course is *W.* by *S.* for the Island *Quicara,* distant 18 Leagues. From *Quicara,* to *Montuosa,* six Leagues, *N. W.* From *Montuosa,* to Point *Burica,* your Course is *N.W.* by *W.* 14 Leagues. Observe that in this Way, four Leagues from *Montuosa,* there are two small Islands call'd *los Frayles,* or the *Fryers:* Four Leagues from *Montuosa,* give the Island to the Landward a *Berth,* for it has a Shoal running a League from it to the *Eastward.* From Point *Burica,* to *Cabo Blanco,* or, White Cape, the Course is *N. W.* and in the Mid-way, that is, 16 Leagues from Point *Burica,* is the Island *del Cano,* where you may wood and water on the *North* Side. These Directions from *Morro de Puercos,* to Cape *Blanco,* are to Seaward, from Point to Point, and from Island to Island; in case you will not put into the Bays, in the following Directions you may see the Land-marks, Depths, and Entrance into Harbours.

If you would coast it along from Point *Mariato,* where the Mountains of *Guanico* terminate, to *Nicaragua,* or *New Spain,* from Point *Mariato,* to the Island of *Sebaco,* is three Leagues, and the Island is about five Leagues in Length, lying *North* and *South,* all wooded, and the new Point of it is about a League and a half from the Continent, with a good Channel between them, through which they pass who are bound for *Philipinas,* if they think fit. At the *North* Point of the Island, is a Shoal; on the *West* Side of it, is the Island *Governadora*; close by, and on the *West* Side of the *Governadora,* is another Channel leading into the same Bay of *Philipinas.* When you sail into this Bay, you are to found all the Way, because there are Banks of Sand, and you may be left aground upon the Ebb, therefore come not into less than six Fathom Water. If you would go up to *Philipinas,* which is a *Spanish* Town, you must anchor to the Leeward of that they call *Isla de Leones,* or, Lions Island, that is on the *North* Side of it. The Channel is close under the Continent;

The South Sea *Coasts.*

tinent; take Heed of coming near the Island, because there are Shoals, and you must anchor at the End of it, a little to the *Westward.* To go up to the River of *Philipinas,* which is two Leagues, you must steer *North* to the Mouth of it, and at the opening is a little high Wood of Mangroves. From the Mouth of the River, to the Town, is three Leagues. There are many more Islands in this Bay, and several Channels; but the best to go out, is that to the *S. E.* of the Island *Governadura.* There is nothing to fear, but what may be seen. There is anchoring on the Outside of it, any where in 15 or 20 Fathom Water.

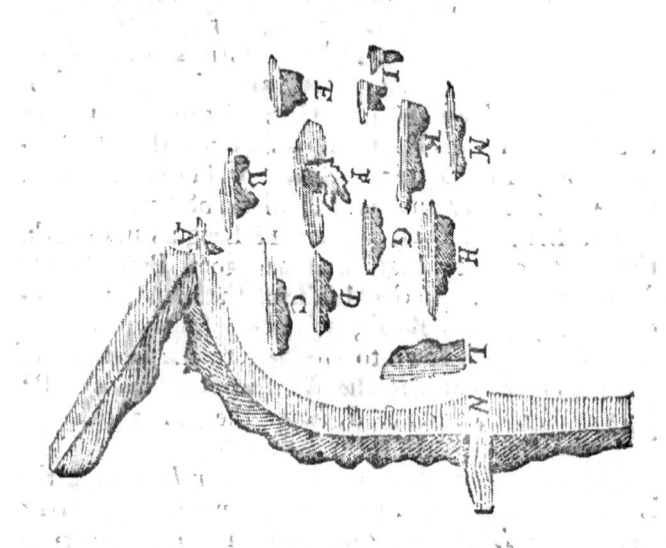

A. Point *Mariato.* B. The Island *Quicara.* C. *Baco* Island. D. *Governadora* Island. E. *Montuosa* Island. F. *Coiba* Island. G. H. Islands without Names. I. *Ladrones* Islands. K. has no Name. L. Anchoring Ground. M. has no Name. N. *Philipinas* River.

To the *Eastward* of *Sebaco* Island, the Continent lies as far as the Island *Canales East* and *West*, and is low Land 'till within four Leagues of the said Island *Canales*, where a little Mountain of copling Land rises next the Coast, very woody. If from the Island *Sebaco*, you would sail for that of *Quicara*, your Course is *S. W.* 15 Leagues.

If from the Island *Quicara*, you would sail for *Pueblo Nuevo*, or, new Town, your Course will be *West* for the Island of *Canales*, which is the same above said to be at the End of the Coast which runs from *Philipinas Westward*. To the *Southward* of the Island *Canales*, is the Island *Corba*; and near *Corba* another small Island, call'd *la Rancheria*, forming a Bay to the *Northward*, opposite to the Head-land, with a sandy Shore. Here is good anchoring, Wood, and Water, and Shelter from the *North* Wind. Upon Occasion you may take Shelter in this Harbour, which is call'd *la Rancheria*, and if you stand in need of Masts or Yards, you will find very good on the Island of *Coiba*.

This Island of *Coiba* is about 18 Leagues in Length, the Land not very high, and has good Ports to the *North* and *West*. In the Middle of the Island, is a sandy Shore, with a Break, whence a Stream of good fresh Water runs down to the Sea, to the Leeward of a sharp Point that is to the *Northward*. This is call'd Port *de Mas*. At the *S. E.* Point there are Shoals and Ridges of Rocks, give them a Berth.

To the *Northward* of the Island *Canales*, where the Land which runs from *Philipinas* terminates, is a Bay call'd *Baia Honda*, or, deep Bay, a very good Harbour, land-lock'd against all Winds, with 15 Fathom Water.

If from the Island *Canales*, or *Baia Honda*, you would go to *Pueblo Nuevo*, take Notice that two Leagues to the *Northward* of the said Island is the said Bay of *Baia Honda*; and thence the Coast runs on to *Chiriqui*, and in the mid Way is the Town of *Pueblo Nuevo*, or, new

new Town, the Diſtance between *Baia Honda* and *Pueblo Nuevo* being ſeven Leagues, the Courſe N. by W. Right before the Mouth of this River, about a League from the Continent, is an Iſland, flat at the Top, round, full of Trees, and ſmall. When you are come up with this Iſland of *Pueblo Nuevo*, the Channel runs on the *Eaſt* Side to the very Mouth of the River 10 or 12 Fathom deep; and the Iſland and the River bear from one another *N. E.* and *S. W.* As ſoon as you come in, you'll ſee a low Point cover'd with Mangroves on the Starboard Side, call'd Point *de la Rancheria*, where Ships uſe to be built; and about a Muſket-ſhot farther up, you may anchor where you pleaſe in 5 Fathom Water. From this anchoring Place, to the *Spaniſh* Town, is 3 Leagues up the River; but the Town may be ſeen from the ſaid anchoring Ground, becauſe it ſtands on an Eminence, to the *Northward*. Ships may go up beyond the Town, tho' there is but little Water. *Note*, That there is no paſſing between the Iſland of *Pueblo Nuevo*, and the Continent, on the *Weſt* Side, becauſe there are many Flats, and the Sea breaks on them.

A. The

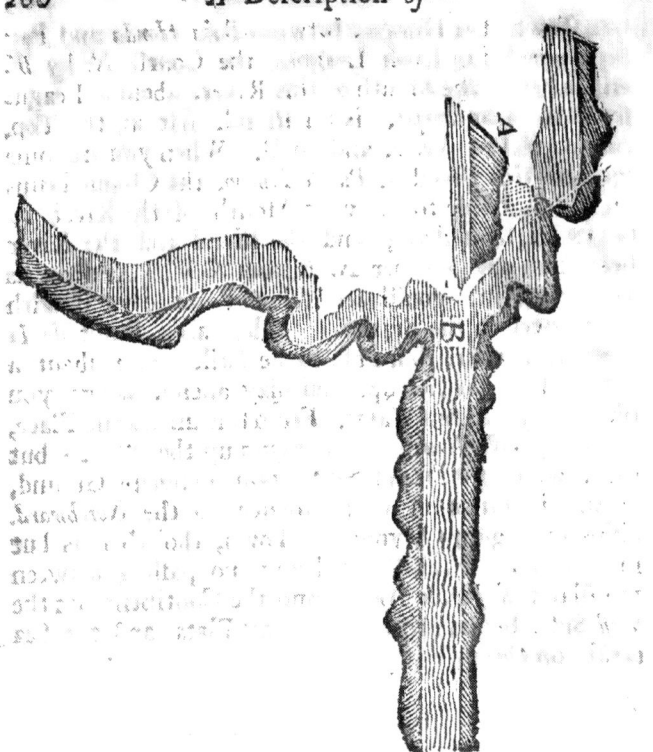

A. The Island of *Pueblo Nuevo*, or, new Town. B. The River of *Pueblo Nuevo*.

As you come out from *Pueblo Nuevo*, towards the S. W. about four Leagues, there are three or four Islands, call'd *Islas de Contreras*, or, the Islands of *Contreras*; and to the *Westward* there are four other Islands, called *Islas Secas*, that is, dry Islands, not but that they have abundance of Trees, particularly Coco Trees; and they all afford Water.

From *Islas Secas*, or, the dry Islands, to *Chiriqui*, four Leagues, and here ends the Coast and Bay, which lies with the Island *Canales East* and *West*. At the
Mouth

Mouth of this River *Chiriqui*, are eight or ten Iſlands, great and small, among them some Shoals the Sea breaks on at low Water. If you are to put into *Quiriqui*, which is a *Spaniſh* Town, you'll see an Iſland close to the Mouth of the River, about a League in Compaſs, and you may go up on either Side of it, for there is Depth enough, and all the Iſlands of *Chiriqui* afford Water and Coco Nuts.

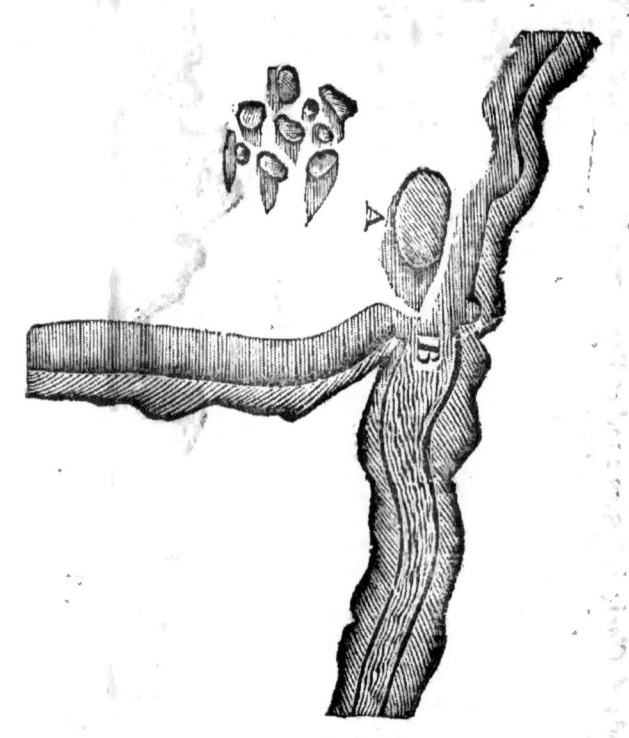

A. The Iſlands of *Chiriqui*. B. The River of *Chiriqui*.

A Description of

At the Islands of *Chirigui*, begins another Bay, extending to Point *Burica*, lying *N. W.* and *S. E.* the Bay running up to the *Northward*. From the Islands of *Chirigui*, to Point *Burica*, is six Leagues. On the *North* Side of this Point *Burica*, within the Bay, is a Port, where Ships may anchor and water. Departing this anchoring Ground, towards the *N. W.* for Point *Burica*, you'll see some Ridges of Rocks; and near the first Ridge to the *Northward*, is Port *Limones*, or Lemons. When you come out from Port *Limones*, give Point *Burica* a Berth; for there are many Flats, running a League out to Sea; and two Leagues to the *Westward* of this Point *Burica*, you'll see the Point of a little white Wood of Mangroves. There is a Port, where the Sailors gather Coco Nuts, when the Trade-Winds blow, which are very frequent; for then the Wind blows upon the Land; but when the opposite Winds reign, they cannot come to gather them, because there is a great Surf.

Port *Limones*. Point *Burica*.

When they bear *N. W.* shew thus.

From

The South Sea *Coasts.*

From Point *Burica,* to *Golfo Dulce,* or, fresh Water Bay, four Leagues, the Coast lies *N. W.* and *S. E.* and within the Bay is a Head-land; on the *N. W.* Side whereof are two little Rocks, near the Continent. When half a League within this Bay, you may anchor, if there be Occasion; for it is a very good Harbour, has fresh Water, and there is nothing to fear, but what is in Sight. All the Coast above-mention'd is high, and very mountainous.

From *Golfo Dulce,* to the Island *del Cano,* seven Leagues, lying *N. W.* and *S. E.* The Island is a League from a very sharp Point on the Continent; and they bear from one another *North* and *South,* forming between them a Bay, in which is another Island, about two Leagues in Compass, inhabited by *Indians*; and from the Island *del Cano,* to this inhabited Island, is about four Leagues.

From the Island in the Bay, to the River *de la Estrella,* or, of the Star, five Leagues *N. W.* and *S. E.*

From the Island *del Cano,* to *la Herradura,* or, the Horse-shoe, 16 Leagues; and this is that they call *Costa Rica,* inhabited by the *Indians,* call'd *Buracos,* and the *Coles,* who are all peaceable, and supply the *Spaniards* who travel by Land.

From the River *de la Estrella,* to this same *Herradura,* 11 Leagues, being a Bay, in which the Sea runs high, the Mouth of it lying *W. S. W.* and on the *West* Side of it is an Island at the Point; the Mouth of the River is about half a League wide.

The

A Description of

The Island at the Mouth of *Rio de la Estrella*,

Shews thus, when bearing *N. N. W.* and as you leave it to the *S. E.* the Peek at (o) opens. In this same Position you'll see towards the *N. W.* Point *Herradura*, which is hilly; and if the Weather be clear, you may see the Mountains of *Costa Rica*, being the highest of all that Coast.

From *la Herradura*, to the Island of *Chira*, 15 Leagues *N. N. W.* and *S. S. E.* and about the mid Way is a *Spanish* Town, call'd *Landecho*, where there are Herds and Flocks of Cattel. The Land along the Coast is low, with many Creeks, and abundance of Mangroves, as far as the River of *Cipanso*, which is two Leagues beyond *Chira*, whither the Ships go to take in the Lading that is brought from *Nicoya*. This Island of *Chira* is inhabited by *Indians*, and affords Water and Provisions. Close by it, is another low round Island; and on the *N. E.* Side of it is a Bank the Sea washes over. *Note*, that to go into *Chira*, you must keep close to the Island, leaving all the other Islands on your Lar-board Side, except the little one where the Shoal is. The Channel lies between the great and the little Islands leading up to the Town, which

which you'll fee by the Sea Side, and there you may water, and be fupply'd with other Neceffaries.

From the Ifland of *Chira*, to that of St. *Luke*, eight Leagues. They bear from one another *N. N. E.* and *S. S. W.* and in the mid Way there are three Iflands, call'd *Iflas de en Medio*, or, the middle Iflands. From *Chira*, to thefe middle Iflands, it is all Shoal, and there is not above fix or feven Fathom Water. Ships failing this Way, muft always keep clofer to the Iflands, than to the Continent, which is low Land, full of Mangroves. Not far from thefe middle Iflands, and neareft to the outwardmoft of them, is another, called the Ifland of *Guayavas*. At the Ifland of St. *Luke*, is a Port, where they lade Mules and other Things for *Panama*. The Harbour is in the loweft Part of the Land, half a League from St. *Luke*'s Ifland, and is call'd *Faro*, where you may water, as alfo in all the Iflands, lying in a Triangle. Towards *Cabo Blanco* there are many fmall Iflands clofe to the Continent.

From *la Herradura*, above-mention'd, to *Cabo Blanco*, or, white Cape, 30 Leagues, lying *Eaft* and *Weft*. Cape *Blanco* and the Ifland *del Cano* bear from one another *N. W.* by *W.* and *S. E.* by *E.* the fame Cape and the Ifland of St. *Luke N. E.* by *N.* and *S. W.* by *S.* diftant nine Leages. Cape *Blanco* is high Land next the Sea, floping away, and then makes as it were a Table up to the Mountain; and from the Offing, looks like an Ifland. Clofe by this Cape is a little Ifland, on the higheft Part whereof is a black Spot of Trees, and it is in 10 Degrees of *North* Latitude.

Cape

272   A Description of

Cape Blanco,

When bearing *N.W.* shews thus; the Land between this Cape and *la Caldera*, lies *North* and *South*; and Point *Guiones* and it bear from one another *N.W.* and *S.E.*

The same bearing *N.N.W.* looks like an Island, thus.

The same, when the Head at the Cross bears *N.E.* distant three or four Leagues, shews thus, and the Land runs away full of Hillocks, to the *N.W.* as far as Point *Guiones*.
From

The South Sea *Coasts.*

From Cape *Blanco*, to Point *Guiones*, 10 Leagues; and in the Mid-way there are two Shoals, running a League out to Sea, being equally distant from the Cape and Point, take heed of them. Point *Guiones*, howsoever it bears from you, shews like a steel Cap, and close by it is a little Island, all about the which there are Shoals; therefore give them a Berth. Between Cape *Blanco* and Point *Guiones*, there are some white Rocks in the Bay.

From Point *Guiones*, to *Morro Hermoso*, or, Beautiful Head-land, eight Leagues, they bear from one another *N. N. W.* and *S. S. E.* a clear Coast; this Head-land is high and upright; when near, you'll see the Sea beat on it.

Morro Hermoso, or, Beautiful Head-land,

Bearing *N. N. W.* shews thus.

From *Morro Hermoso*, to Port *Velas*, seven Leagues; the Coast lies *N. W.* by *N.* and *S. E.* by *S.* Port *Velas* lies up in the lowest Part of the Land, and has fresh Water. Three Leagues before you come to the Harbour, you'll see a little Island, with three or four Rocks by it; and farther on, towards the Port, there are three or four other Rocks, which at a Distance look like Ships under Sail; and therefore the Port is call'd *de Velas*, that is, of Sails; and near the Port, is a Ridge of Rocks a League out at Sea, lying along the Coast, about a League in Length.

From Port *Velas*, to Point St *Catherine*, eight Leagues, the Coast lying *W. N. W.* and *E. S. E.* From Point *Guiones*, to this Point of St. *Catherine*, is 22 Leagues *N. W.* and *S. E.* in a direct Course. Two Leagues out at Sea, from Point St. *Catherine*, is a high Rock, and to the Landward of these, there are two little Islands;

Vol. II.   T   and

and again, within the Point in the Bay towards the *S. E.* there are two more small Islands, the Distance between the former and the latter about a League.

At Point St. *Catherine*, begins the Bay call'd *del Papagayo*, or, of the Parrot, stretching towards the *N. W.* about 16 Leagues; and in the mid'st of it appears a burning Mountain, which is near *Granada*, and call'd of *Bonbacho*, cleft from the Top down to the Bottom in the Nature of a broken Saw; and to the *N. W.* of it, is Port St. *John*, five Leagues distant. The Mark to know this Port by, is a Table, about two Leagues in Length, on a Hill that is not very high, and this is call'd St. *John's* Table.

From Point St. *Catherine*, to St. *John's* Port 15 Leagues, *N. W.* and *S. E.* Take heed how you sail by this Bay *del Papagayo*, especially between *November* and *April*, for then the *North* Winds prevail, and make a high Sea, therefore be sure to keep close under the Land. In the middle Part of this Bay, is a fresh Water-River; but there is always a high Sea, and in the Mouth of Port St. *John*, is good Shelter against the *N. W.* Wind; but the *S. E.* blows right it, and makes a great Sea, and then there is no going ashore, because the Coast is high.

The burning Mountain of *Bonbacho*,

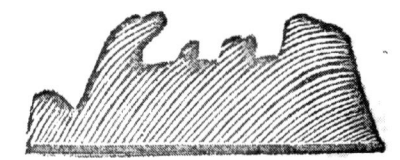

Bearing *N. W.* shews thus.

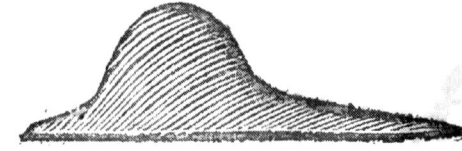

The same bearing *W.* shews thus.

The South Sea Coasts. 275

From Port St. *John*, to *Realejo*, 15 Leagues, *N. W.* and *S. E.* and at this Port St. *John* begin the burning Mountains towards the *N. W.* as far as *Teguantepeque*, being above 200 Leagues along the Coast.

Seven Leagues *N. W.* from St. *John*'s River, is a River of fresh Water call'd of *Tosta*, running down into the Sea, small, and without Mangroves, next the Sea-Coast, and four or five Leagues up the Inland, you'll see a burning Mountain call'd of *Leon*, casting out much Smoke, which is seen at Sea.

The burning Mountain of *Leon*,

On one Side shews thus.

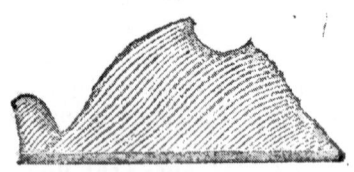

In another Position, thus.

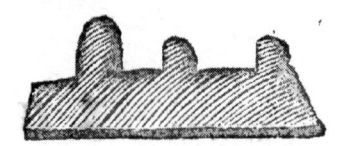

In a third, thus.

From the River of *Tosta*, to *Realejo*, eight Leagues, the Coast lying *N. W.* and *S. E.* and at this River of *Tosta* begins a Ridge of Land, which next the Sea stretches out three Leagues in Length, plain and level.

T 2 The

A Description of
The Ridge of *Tosta*.

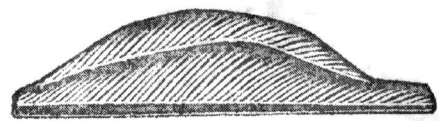

This Land, as you come from the Seaward, shews thus.

The burning Mountain of *Telica*, and this Ridge of *Tosta*, bear *N. E.* and *S. W.* from one another, and the Mountain is six Leagues short of it.

The burning Mountain of *Telica*,

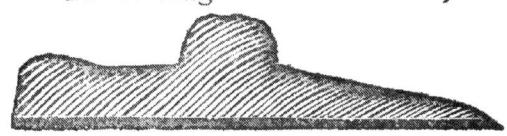

When bearing *N. N. E.* and *E.* shews thus, over the Ridge of *Tosta*.

From the End of the Ridge of *Tosta*, to *Realejo*, is four Leagues, *N. W.* and *S. E.* the Shore low and sandy, full of Mangroves, but deep Water; yet in some Places, as far as *Teguantepeque*, there is anchoring from 15 to 20 Fathom, two Leagues out at Sea, for the most part a sandy Ground. Between the burning Mountain of *Telica*, and that call'd *del Viejo*, or, the old Man's, is a Piece of hilly Land, not very high.

The Land between *Telica* and *Volcan del Viejo*, shews thus.

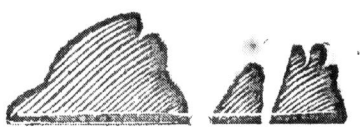

The

The South Sea *Coasts.* 277

Between *Telica* and *Volcan del Viejo*, are two

Other burning Mountains, which shew thus.

From the burning Mountain of *Telica*, to *Volcan el Viejo*, or, the old Man's burning Mountain, six Leagues. This last lies seven Leagues up the Country, bearing with the Bar of *Realejo N. E.* and *S. W.*

Volcan del Viejo,

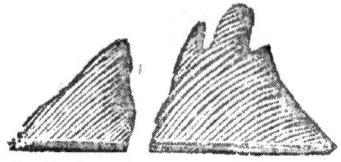

Bearing *East*, shews thus.

The same bearing *West*, shews thus.

There is a great Trade at the Port of *Realejo*, from all Parts along the *South Sea*. If you are to put into it, and stand in from the Seaward, you must come up very close with the Land, to discover the Harbour, which cannot be seen at a Distance, because the Coast is very low Land, and full of Trees; so that unless you come so near as to discern the sandy Strand, which stretches five or six Leagues in Length, there is no seeing of the Port. To know when you are up the Length with this Harbour of *Realejo*, as you stand in from the Seaward, you must bring the burning Mountain above-mention'd, call'd *Volcan del Viejo*, to bear

N. E. then stand in for the Harbour, and you'll soon see the little low Island, which is little above half a League in Compass, being flat at the Top, and cannot be seen 'till within a League. This Island shelters the Harbour, and makes two Mouths or Channels into it; that on the *S. E.* larger than that to the *N. W.* but you must go up the smallest on the *N. W.* Side; for no Ships pass through the other to the *S. E.* because there is little Water, and many Rocks; whereas in that to the *N. W.* there is four or five Fathom Water at low Ebb. When you go in, keep your Starboard Side close up to the very Rocks about the little Island, where the Channel lies, which is narrow, not above half a Cable's Length over. *Note,* That if the *S. W.* Wind should happen to blow hard as you are going over the Bar of *Realejo,* you need not come to an Anchor; for provided there be Day-light enough, and the Wind stiff, you may make your Way at Pleasure, tho' it be Ebb and Spring Tide, there being Water enough. As soon as in, with your Side, as has been said, close up with the Rocks, and that you are come the Length of a Parcel of Spots overgrown with Thistles, which are on the Island; if you are not throughly acquainted with the Harbour, you must come to an Anchor, that at low Water you may see the Flats, which are then bare, and by them you will know the Channel. From the aforesaid Spots of Thistles, strike over to the sandy Point, which lies towards the Larboard Side on the Continent, for that Way the Channel runs. Keep along so, sounding all the Way, in six or seven Fathom Water, 'till you come up to the sandy Point, then run along the Shore next the Point, and drop your Anchor about the mid Way over, for that is the Place to ride. The Town of *Real-jo* is three Leagues up the River. When you go up in your Boat, keep up towards the *N. W.* that is, to the Left Hand, or Larboard Side, 'till you come to a Creek, then hold on the same Way, 'till you come

to

to another Creek farther on; go not up either of them, but proceed 'till you come up with a third, which looks narrower than the others, and that leads up to the Houses. *Volcan del Viejo*, or, the old Man's burning Mountain, above-mention'd, for a Land Mark to find the Port of *Realejo*,

When bearing N. E. shews thus.

A. The

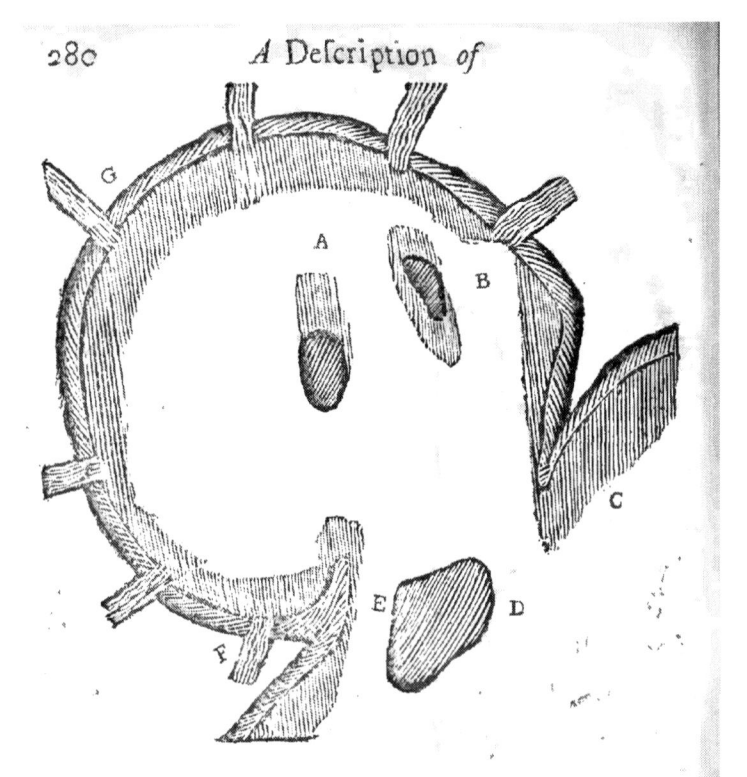

A. The Port of *Realejo*. B. The Island *del Castano*. C. *S. W.* Point. D. *Realejo* Island. E. The Channel into the Port. F. *Aserradores*, or, Sawyers Creek. G. *Realejo* Creek.

From the Bar of *Realejo* to *Confibina*, 12 Leagues, the Coast lying *W. N. W.* and *E. S. E.* When you depart *Realejo*, shape your Course *S. W.* 'till past the Island *de los Aserradores*, or, of Sawyers, which is four Leagues from the Bar of *Realejo*, small and plain, and must be left to the *N. E.* Close to it, on the *S. E.* Side, are two Rocks; and near them is the Mouth of a Creek, call'd *de los Aserradores*, or, of the Sawyers, up which Boats can pass at high Water; from thence

thence the Shoals run out a League to Sea. One League beyond the Island de los Aserradores, towards the N.W. and two Leagues up the Inland, is a round Hill, the Top as it were cut off, and call'd *Mesa de Roldan*, that is, *Orlando's Table*.

Mesa de Roldan,

Bearing N. W. shews thus.

From *Mesa de Roldan*, or *Orlando's Table*, to *la Consina*, seven Leagues, the Coast low and wooded; and about a League up the Country, is a Hill, which bears the Name of *la Consina*. It was formerly a burning Mountain, and burst open, since when the upper Point has remain'd hanging, as is here represented.

A Description of

The burning Mountain of *Confibina*,

Bearing *N. W.* shews thus.

Two other Views of this Mountain are laid down in the Spanish *Manuscript*, without mentioning how they bear, but represented as underneath.

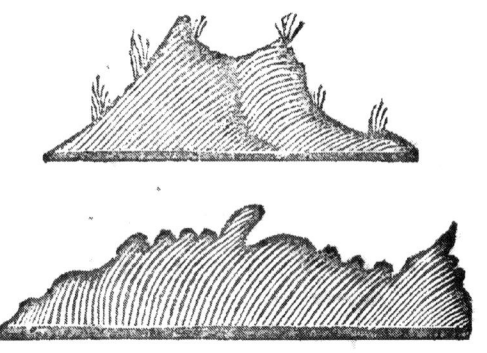

Confibina and *Volcan del Viejo*, or, the old Man's burning Mountain, bear from one another *East* and *West*. At that of *Confibina* there are Docks for building of Ships, and they go thence by Land to *Realejo*.

From the Point and Extremity of *Confibina*, to Port *Martin Lopes*, otherwise call'd *el Condadillo*, or, the little Earldom, eight Leagues; and between them is form'd the Bay of *Fonseca*, running into the Continent; and there is the Ferry they cross who go by Land from *Amapala*, to *Bolcan del Viejo*. This Bay is 10 Leagues over, from *Mapala* to *el Viejo*; and some Ships put into it, to load with Tar, or to careen; and there is every where 10 or 12 Fathom Water.

You

You may go in boldly for what you have Occasion; and as soon as in, you'll see three Islands, stretching out *East* and *West*, two of them large, and the Middlemost smaller. One of them is call'd *la Conchagua*, another *la Miangola*. You must make for *Conchagua*, which lies farthest *West* of the three, near the Continent, sounding all the Way, and come to an Anchor near it, where you think fit. If you are unacquainted, and apprehend any Danger, you may go in your Boat to find out the anchoring Place at *Mapala*, which lies on the *N. W.* Side of the Island, in the Nook of the Bay, which you'll readily find. Take Heed your Cables be good, for the Tides run very strong. These Islands are inhabited by *Indians*, where you may water, and be furnish'd with Masts, and all other Necessaries. Opposite to the Islands, is the River, the Mouth whereof you'll soon see, and on it abundance of Mangroves, fit for any Ship. This River of *Conchagua*, and the Point of *Consibina*, bear from one another *N. W.* and *S. E.* distant eight Leagues. On the *West* Side of this Bay of *Fonseca*, is a high and sharp Hill, call'd the Hill of *Mapala*; and at the Foot of this Hill of *Amapala*, is the Port of *Martin Lopes*, otherwise call'd *el Condadillo*, or, the little Earldom, in large three Degrees Latitude *North*.

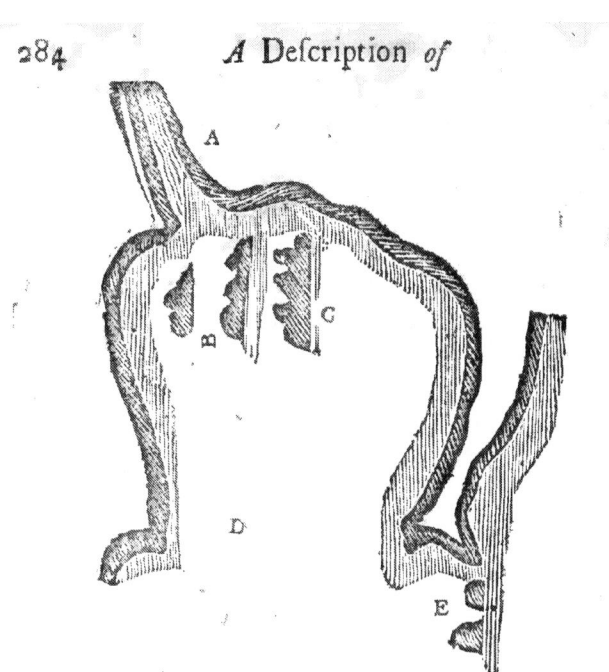

A. *Conchagua* River. B. *Conchagua* Island. C. *Miangola* Island. D. The Bay of *Fonseca*. E. The Rock.

From Port *Martin Lopes*, to the River of St. *Michael*, 18 Leagues, the Coast lying *East* and *West*, high Land next the Sea, and without any Mangroves all the Way to the River, in which there is three Fathom Water at the Flood, so that small Ships may go up it. *N. E.* and *S. W.* with it is a great burning Mountain, call'd of St. *Michael*, lying up the Inland five or six Leagues, in an open Plain, which has no other Eminence about it. This Mountain casts out Smoak, which may be seen out at Sea.

The burning Mountain of St. *Michael*,

Bearing *N. N. E.* at a great Distance, shews thus.

This same burning Mountain is represented in two other Views by the Spanish Manuscript, without mentioning how it then bears; but they are the two next below.

From the River of St. *Michael*, to that of *Guibaltique*, three Leagues *West*, low Land, full of Mangroves next the Sea, with several Creeks. Many Shoals run out about a League to Sea from this River of *Guibaltique*, and the Extremity of it bears with that of St. *Michael N. E.* and *S. W.* and with *Consibina N.* by *W.* and *S.* by *E.* distant 18 Leagues, all the Coast low and shoal, and the Sand Banks lie *East* and *West* along the Coast, two Leagues out at Sea. You may anchor in 15 or 20 Fathom Water, and will see the Country up the Inland is mountainous, and full of Hillocks.

From

From the River of *Guibaltique*, to that of *Linpa*, is all a sandy Shore, with a high Sea continually on the Coast, the Land low, and the Water shoal; so that you may anchor, as has been said, in 20 Fathom.

From the River of *Linpa*, to *Sacatecolula*, four Leagues. This is an *Indian* Town, and some *Spaniards* among them, and the Country about produces abundance of *Cacao*. Near the Town, is a great burning Mountain of the same Name with it, and it is six Leagues from the River *Linpa*.

The burning Mountain of *Sacatecolula*.

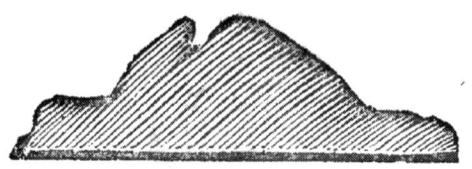

Laid down as in these two Figures by the Spanish Manuscript, but without mentioning how it bears.

Two Leagues up the Inland, and bearing with this burning Mountain N. E. and S. W. is a Hill, like a Sugar-loaf; and at this Hill begins the Coast of *Tonela*, or, of *San Salvador*, that is St. *Saviour*, high Land next the Sea, and the burning Mountain of St. *Saviour* is ten Leagues from that of *Sacatecolula*, appearing over the Coast of *Tonela*, which is an indifferent even Ridge of Land.

The

The high Land between the burning Mountain of
St. *Michael*, above-mention'd,

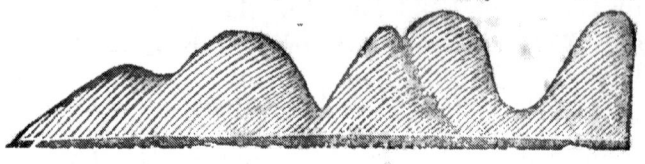

And that of *Sxcatecolula*, appears thus.

The burning Mountain of *San Salvador*,

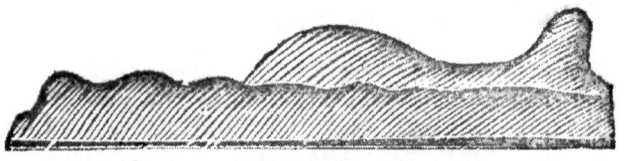

Appears thus over the Ridge of *Tonela*.

Four Leagues to the *Westward* of the burning Mountain of *San Salvador*, is a Hill, near the Coast, call'd *Bernal*; and this and the said burning Mountain bear from one another *E. N. E.* and *W. S. W.*

Bernal.   *Bernalillo*, or, little *Bernal*.

Bearing *E. N. E.* they shew thus.

At this Hill of *Bernal*, commences the Coast call'd *de las Italias*, and terminates the high Land of *Tonela*; and from hence runs the low Land and Coast as far as Point *Remedios*, being ten Leagues, and is call'd the Coast *del Balsamo*, or, of Balsam, being a low Ridge, level at the Top, stretching along the Sea Coast, and terminating before it comes to Point *Remedios*. Opposite to the End of the Ridge, bearing *North* and *South*
with

with it, is an upright Hill, three or four Leagues up the Inland, flat at the Top, call'd the Hill of *Italias*; and it is seven Leagues from the burning Mountain of *San Salvador*, towards the N. W.

From the Hill of *Italias*, to the burning Mountain of *Sonsonate*, is three Leagues; and this burning Mountain bears with Point *Remedios*, where is the Port of *Sonsonate*, North and South.

Point *Remedios* is low next the Sea, and has a Rock standing up by it, about which there are many Sands running out above half a League into the Sea, under Water. If you would go up to anchor in the Port of *Sonsonate*, be sure to give them a sufficient Berth. Point *Remedios* gives Shelter against the S. E. Wind. *Note*, That all this Bay, which makes the Port of *Sonsonate*, is Shoal; and therefore you must sound as you come in, 'till the burning Mountain bears N. by E. giving a Berth to the Point and Sands, 'till you come into twelve Fathom; and when you are in this Depth, and right before the Store-houses, you are in the anchoring Ground. You'll see the Store-houses upon a Crag, and then you'll be half a League from the Land; go no nearer the Shore than 11 Fathom, for there are Mice that gnaw the Cables. If you would ride safe, keep the burning Mountain N. by E. and the Mouth of the River N. E. Be sure to look to your Cables, because of the Mice; and take Heed how you go ashore, for there is generally a great Surf, and you must land near a Parcel of Stones, which are before a Cross in the Nook. From the Port, to the Town of the *Trinity*, is three Leagues. If you would put into this Port of *Sonsonate*, you must, when out at Sea, make the burning Mountain that has the Top broken off, and looks whitish, by Reason of the great Quantity of Ashes about it. Farther on towards the N W. is another little burning Mountain, call'd *de la Paneca*, distant about three Leagues from that of *Sonsonate*.

Near

The South Sea *Coasts.*

Near this last, there are three or four little Sugar-loaf Hills.

The burning Mountain of *Sonsonate*,

Bearing from *N. W.* to *N. E.* shews thus.

The same bearing *S. E.* shews thus.

290　　*A* Description *of*

The Port of *Sonsonate.*

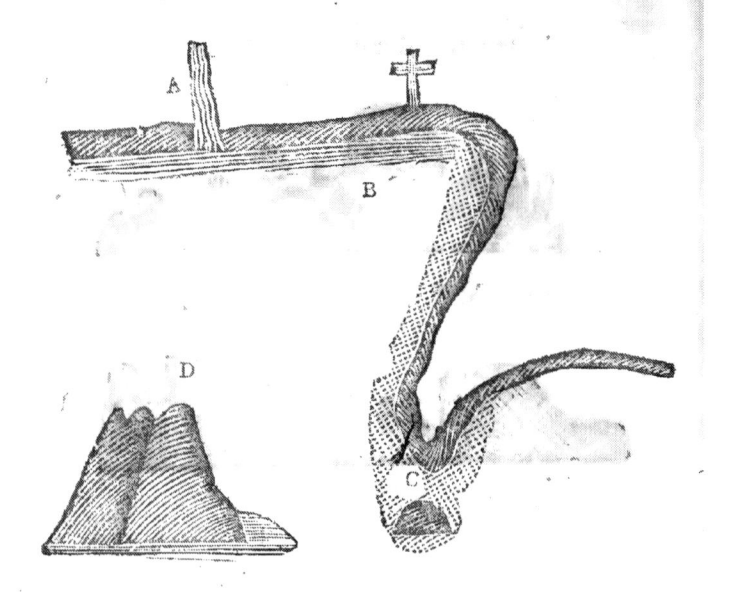

A. *Sonsonate* River.　B. The Anchoring Ground.
C. Point *Remedios.*　D. *Paneca* Rocks.

To ride safe in this Harbour, you must keep the Ridge of *Paneca,* N. by E. and S. by W. in seven Fathom Water, and the Mouth of the River *North* and *South,* distant a quarter of a League.

Between the Port of *Sonsonate,* which others call the Strand of *Catulta,* and the Strand of *Estapa,* the Coast lies *W.* by *N.* and *E.* by *S.* the Distance 26 Leagues. This is the Port of *Guatimala,* all the Coast low Land, sandy, and full of Mangroves next the Sea, and there is clean anchoring Ground all the Way.

From *Catulta,* or *Sonsonate,* to the River *Moticalco,* seven Leagues. This River is much infested with the *North* Wind, and opposite to it there are some small
high

The South Sea *Coasts.*

high Hills, by which it is known; and at the Mouth there are no Mangroves, as there are all along the Coast, except only at this Place.

The burning Mountain of *Moticalco*,

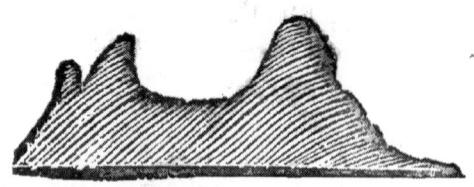

Bearing S. E. shews thus.

Four Leagues to the *Westward* of *Moticalco*, is another great River, which has two Fathom Water upon the Bar. From this River, to the Bar of *Estapa*, is 15 Leagues. This River of *Estapa*, and the burning Mountains of *Guatimala*, bear from one another N. E. and S. W. Note, that the Land-marks to know this River of *Estapa* by, are the tallest Mangroves of the whole Coast growing about it, and in the mid'st of these mighty Trees, is the Bar or Entrance into it.

The burning Mountains of *Guatimala*,

Bearing from *East* to *North*, shew thus.

The same in another Position, is represented thus in the *Spanish* Manuscript; but the Manner of bearing not set down.

The Bar of *Eftapa* and *Teguantepeque*, bear from one another *N. W.* and *S. E.* low Land, and full of Mangroves next the Sea, but high, and with many burning Mountains up the Country; and from the Bar of *Eftapa*, to *Teguantepeque*, which others call *Sequetepeque*, is 65 Leagues, as particularly mention'd below.

The burning Mountain of *Teguantepeque*, or *Sequetepeque*,

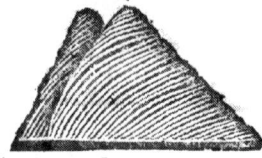

Bearing *N. W.* shews thus.

About 12 Leagues from *Eftapa*, towards the *N. W.* is another burning Mountain, and much high Land, and in the highest Part of all, is this burning Mountain, call'd of *Sapotitlan*.

The burning Mountain of *Sapotitlan*,

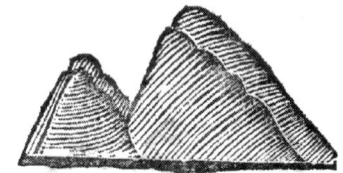

Bearing *N. W.* shews thus.

The same in another Position reprefented, but not nam'd in the *Spanish* Manufcript.

From *Sapotitlan*, to *Milpas*, being two other burning Mountains, twelve Leagues.

The burning Mountains of *Milpas*,

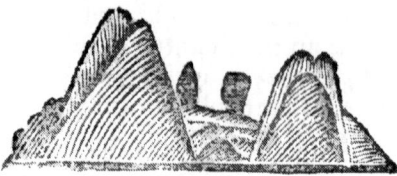

Bearing *N. W.* shew thus:

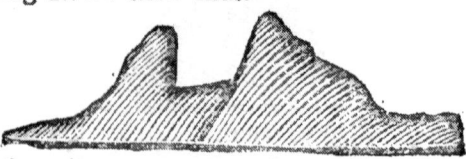

The same in another Position.

From these burning Mountains of *Milpas*, to that of *Soconusco*, 12 Leagues. This is a low burning Mountain, lying near the Sea-Coast, which is also low there.

The burning Mountain of *Soconusco*,

Bearing *N. W.* shews thus.

From the burning Mountain of *Soconusco*, to the *Encomienda*, or Cross, seven Leagues. This Hill of the Cross is low, distant from the Sea about half a League, and on it is a Cross form'd by the Greens growing on the Flat, which is to be seen the Year about, and therefore it is call'd *la Encomienda*, or, the Cross. There is good anchoring all along this Coast; and at this Hill of the Cross Ships take Shelter, when the *North* Wind is boisterous, or else at *Teguantepeque*, if they cannot reach to *Bernal*, or Port *Mosquitos*.

294 A Description of

The Cross above-mention'd,

From the Seaward, shews thus.

From *La Encomienda*, or, the Hill of the Cross, to *Bernal*, which is the Entrance into the Bay of *Teguantepeque*, five Leagues. The Mountain running out from the Inland, towards *Bernal* and the Sea, 'till within half a League of it, as you come coasting from *Soconusco*, this Point seems to run out into the Sea, and the high Land to terminate there, which it does not, but returns up the Inland.

Bernal, at the Mouth of the Bay of *Teguantepeque*.

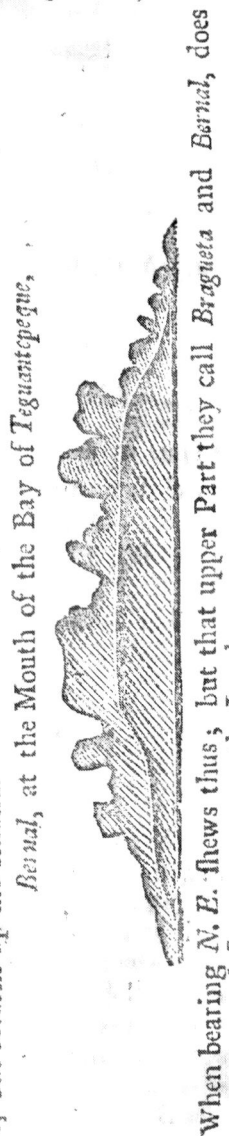

When bearing *N. E.* shews thus; but that upper Part they call *Bragueta* and *Bernal*, does not appear, unless you are near the Land.

The

The South Sea *Coasts.* 295

The same bearing *South*, shews thus.

This Hill of *Bernal*, and the River of *Samitlan*, bear from one another *East* and *West*, distant 41 Leagues, as follows.

From *Bernal*, to Port *Mosquitos*, nine Leagues, low Land, wooded with Mangroves, and subject to be flooded. In this Port *Mosquitos* Ships anchor when the *North* Winds rage; and from it runs a River, which puts out many Banks of Sand towards the *N. W.* and there is a great Surf on them, tho' the Sea runs not very high. All this Country about being liable to Inundations, three Leagues up the Inland, there are some higher Grounds, which look like Islands.

From Port *Mosquitos*, to the Bar and Port of *Teguantepeque*, eight Leagues; and near the Bar there is a Parcel of lofty Hillocks, being Islands in the River, by which it is known where the Bar lies. To the *N. W.* from this River, is a round Hill or Head-land, call'd *Morro del Carbon*, or, Coal Head-land.

Morro del Carbon,

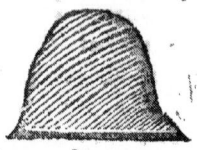

Bearing *N. W.* shews thus.

From the Bar of *Teguantepeque*, to *la Ventosa*, four Leagues. This is call'd *la Ventosa*, that is, the Windy Place, because the *North* Wind is there very boisterous. Here Ships take in the Cannon, and other Stores, which are brought along the River of *Guasacalco*, from the *North* Sea, there being but nine Leagues of Land
U 4 Carriage,

Carriage, to the Ships of *Philipinas*. The Ships that are to cross this Bay of *Teguantepeque*, usually come to an Anchor, as has been said, at *Bernal*, Port *Mosquitos*, or any other about the Bay. When you are to cross this Bay of *Teguantepeque*, be sure to keep as close under the Land as you can; for the farther you are out at Sea, the worse it is; and contend not with the *North* Wind, for Ships are frequently lost by so doing, or at least oblig'd to throw over Part of their Lading. When you are half way over the Bay towards the *N. W.* the Wind comes larger, and you may steer due *West*. You are to be two Leagues from the Land of *Bernal*, when you cross this Bay, and always keep up towards the Land of the Bay.

Morro del Carbon,

In some Part of your Passage, will shew thus.

From the Bar of *Teguantepeque*, to *las Salinas*, or, the Salt-Pits, six Leagues, the Coast lying *E. S. E.* and *W. N. W.* At these Salt-Pits the low Land terminates, and before them are two high Rocks, to the Landward of which is a Harbour for Ships. Hence they carry Salt to *Teguantepeque*.

From *Salinas*, or, the Salt-Pits, to *Puerto de los Angeles*, or, Port Angels, 38 Leagues, the Coast lying *W.* by *N.* and *E.* by *S.* From *Salinas* to *Guatulco*, 25 Leagues, all clean anchoring Ground.

From *Salinas*, to the Head-land of *Macatan*, two Leagues. At this Head-land a fresh Water River falls into the Sea. The Head-land it self next the Sea, is black, very rocky, but the Top of it is white, having a Spot of Sand; and there are no other Spots of white Sand all along the Coast, except two or three at *Salinas*, or, the Salt-Pits; and on the Top of all

the

the Headland is a Peek, like a little burning Mountain, which appears out at Sea, and near the Headland is a dangerous Bank of Sand.

From the Head-land of *Macatan*, to that of *Banba*, four Leagues; and close by it is a dangerous Shoal, a League out at Sea; and they bear from one another *North* and *South*. It is all high Land and Rocks, and next the Sea sandy Strands at certain Distances.

From the Head-land of *Banba*, to the Island of *Itata*, three Leagues. This Island of *Itata* is close up by the Land, within a quarter of a League; and in that Space is good anchoring, shelter'd from all Winds. The Island is small and white, and is cleft in the Middle. Half a League from the Head-land, is a fresh Water River, and an *Indian* Town. The Coast is bold, and when the *North* Wind blows, you may land.

From the Island of *Itata*, to Point *Artea*, seven Leagues, most of it sandy Strands; and between the Island and the Point, is the Town of *Gama*, a League up the Inland. Short of Point *Artea*, is a Farm of black Cattel, where is also Water, and other Necessaries, if you have Occasion. You may safely land at this Point of *Artea*, unless the *S. E.* Wind blow, for that makes a great Surf. This Point is low, and at a Distance looks like a little Island. It affords Shelter against the *N. W. West*, and *S. W.* Winds. All this Coast lies *W.* by *S.*

From Point *Artea*, to the River of *Samitlan*, four Leagues, where is an *Indian* Town, fresh Water, and what else you may stand in need of. Going along the Shore, where the Strand ends, is the River, where you may water, for there is no great Sea.

From the River of *Samitlan*, to the River of *Copalita*, two Leagues. This is a great and deep River, and along it runs the Road from *Guahaca* to the Sea. There is anchoring Ground all along this Coast, the Land clean and level.

To

To the *Westward* of this River of *Copalita*, is the Harbour of *Guatulco*, the Port to *Mexico* and *Guaxaca*, a safe Place, and shelter'd against all Winds, except the S. E. The Mark to know this Port by, for it is close hid up, is a League before you come to it, a little round mountainous Island, call'd *Tangolotango*. At the Mouth of the Harbour, is another little Island, without any Trees. A little farther to the *Westward*, is *el Enfadero*, that is, the Roaring Place; and when the S. W. Wind blows, you'll hear it roar. If you come in from the Seaward, you'll see a high Mountain, make for the Harbour, keeping the highest Part of it right before you. When in *Guatulco*, keep to the *East* Side, for there is most Water, and best Shelter.

A Strand runs from the Port of *Guatulco*, to *Calluta*, whither the Boats go for Water. You'll see a little Island to the *Westward*, where you may anchor to water. About a League to the *Westward* of this watering Island, is another Island; and to the Landward of it is good anchoring Ground, safe from all Winds, it is call'd *Sacrificios*, or, the Island of Sacrifices, and there also you may water.

From this Island of Sacrifices, to the River of *Coiula*, three Leagues, a deep Coast, and bad landing.

From the River *Coinla*, to the River of *Julian Carrasco*, four Leagues. This River runs out at the Strand, where there is a good landing Place; but before you come to it, there is a Bank of Sand, which appears above the Water, like a Tortoise, about a quarter of a League from the Continent, and half a League from the River, bearing from one another N. W. and S. E.

From the River of *Julian Carrasco*, to *Puerto de Angeles*, or, Port Angels, five Leagues; and two Leagues short of Port Angels, is a Creek, which affords very good Shelter; and to the S. E. from this Creek, there runs out into the Sea a Row of little high Rocks, abour half a League, but there is no fresh Water in the Creek. In the Way from it, short of Port Angels, is a dangerous Bank of Sand, under Water; and near it is a little high Rock, which bears with the Port *N. W.* and *S. E.* Port Angels is a good Harbour, and within it, on one Side, is a Creek; the *S. E.* Side affords better Shelter than the other. It is high Land on both Sides. Towards the *S. E.* at the Mouth of it, is a high Rock; and up some Way in the Creek, a Brook runs down to the sandy Shore, and is lost in the deep Sand. A little higher you may see the Brook run down, and a Ground of Reeds. In this Port there is some Wood.

From Port Angels, to the River that runs by *la Galera*, three Leagues; and here terminates the Coast, bearing *E. S. E.* and *W. N. W.* From this River *de la Galera*, runs a large sandy Shore, and above that Shore there are abundance of Mangroves, which may serve to make Masts for Ships, and they extend about two Leagues. The Coast is upright, with anchoring Grounds all the Way, Hills and Dales, with greater and lesser Strands.

From the River *de la Galera*, to the River of *Masia*, 18 Leagues. This is a fresh Water River, swells much in rainy Weather, and sometimes floods the Town call'd *Masia*. Half a League without the River, is a small Island, and a Parcel of little Rocks.

From the River *Masia*, to a Point to the *Westward*, two Leagues. Before this Point, which is low, is a little Rock, and behind the said Point a little Rivulet runs into the Sea. When the Wind blows at *N. W.* you may go ashore for Water, which is to be had to the *S. E.* of the Point, where the Sea is still.

From this Point and River, to *Puerto Escondido*, or, hidden Harbour, eight Leagues. *Puerto Escondido* is a small Bay, having a Point which runs out into the Sea half a League beyond the Shore; and a little Way within the Point, is a small Island, which closes and makes the Harbour. There is good anchoring to the Landward of that little Island, tho' the *W.* and *S. W.*

W. Winds drive in ; however, you may safely go a-shore.

From *Puerto Escondido*, or, hidden Harbour, to *Pesqueria de Don Garcia*, or, *Don Garcia*'s Fishery, 30 Leagues, the Land all Vales, and open Strands, without any Harbour. Ten Leagues to the *Westward* of *Puerto Escondido*, before a Parcel of Crags, there are many dangerous Shoals, which run out two Leagues into the Sea, and shew the Bottom ; for there is but a Fathom, or a Fathom and a half Water at low Ebb ; take Heed of them, for they lie above two Leagues in Length ; the Crags and they bear from one another *N.* by *E.* and *S.* by *W.* distant two Leagues.

From these Crags and Shoals, to the Port of *Acapulco*, 25 Leagues, the Coast lying *W. N. W.* and *E. S. E.* Three Leagues to the *Westward* of the Shoals, is a little white Island, where there is anchoring Ground, and is call'd the Island of *Alcatrases*, being half a League from the Continent. To the *Westward* of the said Island, is a River, which runs out as far as the Island, and there Ships may water. The Coast is mountainous, and the Shore upright, and between this Place and *Acapulco* there are abundance of Crags.

From this River to *Don Garcia*'s Fisheries, 15 Leagues, a spacious Shore, stretching out as far as Port *Marquis*, which others call *Puerto Chico*, or the little Port. Near these Fisheries there are two little Rocks, and a fresh Water River ; the Place generally very still, but when the Sea swells, it is boisterous.

Between these Rocks and the Port of *Acapulco*, the Coast bears *N. W.* by *W.* and *S. E.* by *E.* to Port *Marquis*, 5 Leagues, and from Port *Marquis*, to *Acapulco*, one League. If you come in from the Seaward, you'll see four Mountains, the first next the Sea somewhat high, and the others rise higher gradually behind one another, and on the highest, is a burning Mountain toward the *S. E.* At the Foot of all these Mountains, is the Harbour of *Acapulco*, spacious and very safe ; and a little without the Mouth of it is an Island. On the *N. W.* Side of this Island, is a narrow Channel, at which Ships may safely go in; for there is Water enough. When you sail in on the *S. E.* Side, which is a wide Channel, you'll see a Bank of Sand above Water, call'd *el Grifo*, leave it on your Larboard-side; but you must keep close to it to sail in, and very hard up with a little Point there is by it, and you'll soon see the Town within. To the *Westward*, on the Shore, you'll see two little Rocks. Port *Marquis* stretches out *N. W.* and *S. E.* a League. The Mouth of it lies *E. S. E.* and *W. N. W.* It is a safe Harbour, and very deep, having 20 Fathom Water. The Marks to know it by, are high Land, rocky, and next the Sea is a round Hill somewhat higher. Opposite to the Harbour, on the said Hills, you'll see some red and white Stones, which appear, as it were, through a Mist. Observe the Landmarks well, for the Harbour lies up very close; take special heed that you miss not the Island. On both Sides of Port *Marquis*, as far as *Acapulco*, is rocky for above two Leagues, and on both Sides of those Rocks low Land and sandy Shores, and in the midst of those Rocks is the Port of *Acapulco*. Note the Land-marks carefully.

The Mountains of *Acapulco*.

These are the four Mountains that appear above the Port of *Acapulco*, and the burning Mountain is towards the *S. E.* of the highest, the lowest is next the Sea.

A Description of

The Island of *Acapulco*, Bearing *N. N. W.* shews thus.

A. The

The Port of *Acapulco*.

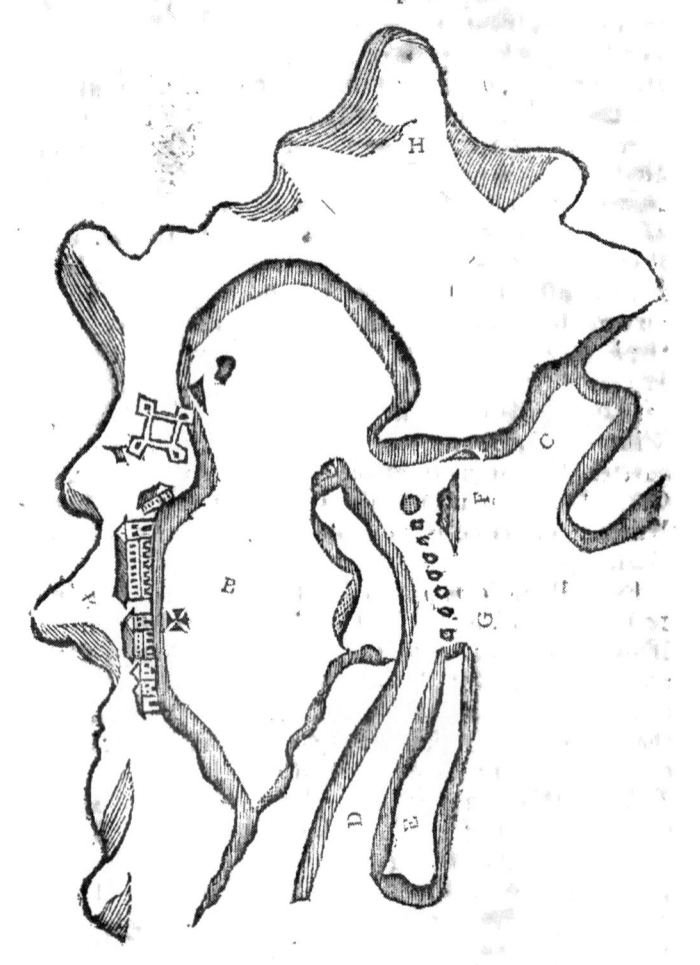

A. The Town of *Acapulco*. B. The Harbour. C. Port *Marques*. D. The little Channel. E. The Island. F. The little Island, call'd *el Grifo*. G. The great Channel. H. The Mountains call'd *Cerros de la Brea*, or, Tar-Hills.

From

From the Port of *Acapulco*, to that of the *Nativity*, 70 Leagues, the two Ports bear from one another N. W. and S. E. Coming out of the Port of *Acapulco*, to the *Westward*, you'll see a spacious Strand, extending above 24 Leagues, all of it low next the Sea, with many Palm-Trees in several Places, and is call'd the Strand of *Sitala*, or of *Apusaguale*. At 18 Leagues distance from *Acapulco*, you'll see a Spot of Mangroves higher than all the other Trees, about a League up the Inland, and stretching out half a Leagues along the Coast, which is here call'd *Tequepa*. Five Leagues farther N. W. is an upright Point next the Sea, not very lofty, and the highest Part of it, at a Distance, looks like Islands. Here is Shelter against the *West*, the S. W. and the *South* Winds, which are the most boisterous on that Coast in Winter. You'll also see a white Rock, standing out a quarter of a League from the Land; there is anchoring Ground between it and the Continent in 10 Fathom Water. The same is all along the Coast, the Bottom clean.

From Point *Tequepa*, to the Head-land of *Petaplan*, 10 Leagues, N. W. This Head-land looks like a little Island, and a quarter of a League to the Seaward of it, are three very white Rocks, which at a Distance look as if they were all but one. You may pass between them and the Head-land, and come to an Anchor close to them, next the Continent, in a convenient Depth. There is Shelter against the *South* and S. W. Winds, on the N. E. Side of the Head-land, because the Coast stretches out *East* two Leagues. It is all clean, and you may land upon the Strand behind the Head-land. At the End of the Bay, about half a League up, there are *Spaniards*, and an *Indian* Town.

About four Leagues N. W. from *Petaplan*, is a little Rock, half a League from the Continent, the Coast lying *North* and *South*. About the Length of these Rocks, is a good Harbour, call'd *Siguatanejo*. Note, That tho' this Port cannot be seen from the Offing, as
soon

soon as the Rock comes to bear *North* you'll see it, and may pass by either Side of it. A League farther, towards the *N.W.* there are five or six Islands, great and small, where is a Village, but inconsiderable, towards the *S.E.*

Two Leagues still *N.W.* from these Rocks, is a Spot of high hilly Land, call'd the Land of *Tacomatlan*; and before this Land, close to the Sea, is a Town call'd *Istapa*; and on the Brink of the Sea, is a small Spot of Land, which looks like an Island, shelter'd from *East* to *S.E.* and this Spot of Land may be seen at above ten Leagues distance every Way, because it is high, and the rest of the Land low.

To the *N.W.* of *Istapa*, is a flat level Shore, without any Harbour for about 12 Leagues, in some Places full of Trees, and at the End of it a Spot of thick and green Mangroves. There is the Mouth of a large River, call'd of *Sacatula*, and half a League up the River, is a *Spanish* Town call'd the Town of *Sacatula*. Note, That you must keep within two or three Leagues of the Land, to be able to make these Land-marks. Over this River of *Sacatula*, next the Sea, are some Hills, the least of them open without Trees.

From this River of *Sacatula*, the Land runs away *N.W.* rugged next the Sea, full of Hillocks, of a moderate Height, and is call'd *los Motines*. This high rugged Land holds for 25 Leagues. In the highest Part of this Land, about half a League beyond *Sacatula*, you'll see two, as it were little Dugs, very close together; and when you are near the Land, bearing *North* and *South* with them, you'll discover an indifferent high Rock, with a Spot of Strand like a Creek; when you are posited *North* and *South* with it, you may discern the white Church of a great Town call'd *Tutapan*. You may anchor to the *Southward*, between that Rock and the Strand, in four Fathom Water, clean Ground; and if you would go ashore, make up to the End of the Strand, towards the *N.W.* near the Stones, and

Vol. II.   X   you'll

you'll presently see the Way to the Town, which is inhabited by *Indians*.

Four Leagues to the *N.W.* of *Tutapan*, is a Point indifferent high, with a Parcel of Rocks by it on the *S.E.* Side, which you will not see, unless near the Land. Between these Rocks and the Land to the *S.E.* is a Piece of a Strand, like a Creek, and a very green Vale. Here is good Anchoring and Shelter from the *West* and *N.W.* Winds, in 12 Fathom Water, and the Place is call'd *Muibata*. If you have Occasion to go ashore, you'll find *Indians*, who generally reside there, and follow Tillage. There you'll see a River, which runs only in Winter, and the Way, which leads up the same River to the Town, standing on the Top of a Hill call'd *Pomaro*.

Six Leagues *N.W.* from this Town of *Pomaro*, is a high Point perpendicular next the Sea, looking like an Island, or small Head-land, call'd *Tuchifi*; and here ends the Land above-mention'd call'd *Motines*. Tho' this be rugged Land, like all the rest of the Coast, yet there are Strands, and anchoring Places, and Shelter from the *N.W.* Wind, which is the most boisterous along that Coast during the Summer Season. To the *N.W.* of this Point, is plain Land full of Mangroves, keep an Offing of three Leagues from it. From this Point you'll see a Parcel of Ridges or high Land, all champion Country, which is call'd *Colima*. Among these Ridges, is a smooth Break running *N.E.* up the Country; and, if it be clear Weather, you may discern at the farther Part of the Break a burning Mountain, continually smoaking, and call'd the burning Mountain of *Colima*. It is all cover'd with Cattel, and Orchards of *Casao*.

Eight Leagues from the Vale of *Colima*, is a very round Head-land call'd *Santiago*; and on the *S.E.* Side of it, are two Hillocks, like Dugs; between those Dugs and the Head-land of *Santiago*, is the Port of *Salagua*. In order to know this Port of *Salagua*, you are to obferve, that there is a very white Rock clinging clofe to the Head-land of *Santiago*, which may be feen at eight Leagues Diftance, whatfoever Way you come towards it. Between this Rock and the oppofite Point, being about three Leagues Diftance, is a Bay with a Strand, and farther up it is all wooded. If you would put into this Port of *Salagua*, ftand ftrait in for the Strand, for at the Ends of it, there are two very good Harbours, where many Ships may ride. They are call'd *las Calletas*, that is, the Creeks; and that which is to the *N. W.* of the faid Strand, is alfo very fafe, Land-lock'd againft all Winds, tho' fmaller than the other. In this Port of *Salagua*, is a frefh Water River, and there are Plantans and Wood. As foon as landed, you'll fee the Road that leads to *Salagua*, which is a League and a half from the Sea. Note, That between *Salagua* and the white Rock, is the Port of *Santiago*.

Six Leagues *N. W.* from the white Rock, is a little Head-land, which afar off looks like an Ifland, and when near, will appear to be an indifferent Head-land, all craggy next the Sea, with a little Rock clofe by it, made like a Sugar-loaf very fhapable. On the *N. W.* Side of this Rock, is a Strand about a League in Length, call'd the Port of the *Nativity*.

At the End of the Coaft, which forms the Port of the *Nativity*, towards the *N. W.* is another Port, by the Natives call'd *Melaque*, Land-lock'd againft the *N. W.* the *Weft*, and the *S. W.* Winds.

Bare three Leagues from Port *Melaque*, is a Row of four or five Rocks above Water, or fmall naked Iflands, running from the Continent, and ftretching out *N. E.* and *S. W.* and if the Weather be fair, you'll fee the burning Mountain of *Colima* to the *Eaftward* up the Country, fmoaking. The Coaft between thefe Rocks and the Port of *Acapulco*, lies *E. S. E.* and *W. N. W.*

Four Leagues *N. W.* from thefe Rocks, or fmall Iflands, are two other fuch Rocks or Iflands, about half a League from the Shore, and, at a Diftance, look like

like Ships under Sail, call'd the Rocks of *Aquiapafulco*. You may fafely anchor near them, clofe under the Shore, for Shelter againft the Sea and Wind.

Between two and three Leagues to the N. W. from thefe Rocks, is a low Point, with red Crags, and a little Rock or bare Ifland clofe to it, on the N. W. Side whereof is good anchoring, under Shelter from the S. E. to the S. W. At this Point runs in a Bay trending towards the N. Weft, about eight Leagues, where you'll fee two or three fmall low Iflands, call'd the Iflands of *Chametla*; between which and the Continent is very good anchoring. The Way in, is from the S. E. and there is a Fifhery belonging to the Town of the *Purification*, which lies fourteen Leagues up the Country.

From thefe Iflands of *Chametla*, the Coaft runs to the N.W. a ftrait Shore, as far as Cape *Corrientes*, or Currents. When near the Cape, if you happen to meet with any Squals of Wind at N. W. there is a Parcel of upright white Crags next the Sea, make directly for them, becaufe to the S. E. clofe up there is very good anchoring, fhelter'd from the N. W. the *Weft*, and the S. W. Winds, the Place call'd *las Salinas del Piloto*, or, the Pilot's Salt-Pits, by Reafon Salt is made very near this Port. The aforefaid Cape *Corrientes*, or Currents, being in 20 Degrees of *North Latitude*, is indifferent high Land rifing by Degrees, barren and bearing few Trees; but up the Country there appears above it a high Ridge of Mountains, forming many Heads, and call'd *los Coronados*.

From Cape *Corrientes*, there runs in a Bay E. by S. 10 or 12 Leagues. All the Land, to the N. E. and N. N. E. is low, and looks very pleafant to the Eye. This Bay is fix or feven Leagues in Breadth, and all the low Land, which is full of *Guayavas*, Cacao, and Breeds of Mules, belongs to the Liberty of the City of *Compoftela*.

From Cape *Corrientes*, or, Currents, to the Point at the other End of the aforesaid Vale, is about 10 Leagues, N. by E. and S. by W. That Point forms a round Head-land, of an indifferent Size, which looks like an Island, without any Trees, and is call'd Point *Ponteque*. In the Offing, to the *Westward* of it, are two small Islands call'd the Isles of *Ponteque*, almost a League from the Continent. Ships may safely pass between them and the Shore. On the *West* of these Islands, are some small white Rocks, and then a Bank of Sand on which the Sea breaks, at the End whereof are two other little Rocks, the whole extending two Leagues.

Three Leagues to Seaward of these Rocks, is another small one, clove in the middle, which, at a Distance, looks like a Ship without Masts. You may safely pass between this and the Rock of *Ponteque*.

About 14 Leagues N. W. by W. from the said Rock, are three large Islands and a small one, the three greater call'd *las Tres Marias*, or, the Three *Marys*; and the lesser, *la Isla Baxa*, or, the Low Island, lying towards the N. E. and by it a white round Rock; all these Islands lie N. W. and S. E.

From the Rocks of *Ponteque*, the Coast runs N. E. above 20 Leagues, to the Port of *Matanchel*; and if the Weather be clear, you'll see a very high Hill over the Port, with a Break on the Top, which is call'd the Hill of *Xalisco*, and may be very well made eight or nine Leagues before you come to the Port of *Matanchel*. In a Bay of sandy Shore, joining to some high Land, at half a League distance from the Shore, you'll see a small, round, mountainous Island, call'd *Maxantella*, and on the Shore opposite to it, are Orchards of *Cacao*, and grazing Lands. About two Leagues to the N. W. of this Island, is a Piece of Land full of small red Crags, where this Course ends.

None of the *Spanish* Manuscripts which I have seen, go any farther *Northward* in the Description, nor do they afford us any Draughts beyond *Acapulco*.

A Description of

Courses and Distances.

	Leagues.
From *Panama*, to Port *Perico*, S. W.	2
From *Port Perico*, to *Otoque*, S. W.	4
From *Otoque*, to the Island *Iguanas*, S. W.	2
From the Island *Iguanas*, to Point *Mala*, S. W.	2
From Point *Mala*, to Point *Hignera*, N. W.	7
From Point *Hignera*, to *Morro de Puercos*, W. by N.	2
From *Morro de Puercos*, to Point *Mariato*, *West*,	12
From Point *Mariato*, to the Island of *Sebaco*, *West*,	3
From the Island of *Sebaco*, to that of *Quicara*, S. W.	15
From the Island *Quicara*, to *Baia Honda*, or, deep Bay, *West*,	10
From *Baia Honda*, to *Pueblo Nuevo*, or, New Town, N. by W.	7
From *Pueblo Nuevo*, to the Islands of *Contreras*, S. W.	4
From the Islands of *Contreras*, to *Islas Secas*, or, the dry Islands, *West*,	1
From *Islas Secas*, to *Chiriqui*, N. W.	4
From the Islands of *Chiriqui*, to Point *Burica*, N. W.	6
From Point *Burica*, to *Golfo Dulce*, or, fresh Water Bay, N. W.	4
From *Golfo Dulce*, to the Island *del Cano*, N. W.	7
From the Island *del Cano*, to the Island in the Bay, N. W.	4
From the Island in the Bay, to the River *de la Estrella*, or, of the Star, N. W.	5
From the River *de la Estrella*, to *Herradura*, or, the Horse-shoe, N. W.	11
From *la Herradura*, to the Island of *Chira*, N. N. W.	15
From the Island *Chira*, to that of St. *Luke*, N. N. E.	8
From *Herradura*, above-mention'd, to *Cabo Blanco*, or, white Cape, *West*,	30
From Cape *Blanco*, to Point *Guiones*, *West*,	10

From

The South Sea Coasts.

From Point *Guiones*, to *Morro Hermoso*, or, beautiful Head-land, N.N.W. 8
From *Morro Hermoso*, to Port *Velas*, or, Sails, N.W. by N. 7
From Port *Velas*, to Point St. *Catherine*, W.N.W. 8
From Point St. *Catherine*, to Port St. *John*, N.W. 15
From Port St. *John*, to the River of *Tosta*, N.W. 7
From the River of *Tosta*, to *Realejo*, N.W. 8
From *Realejo*, to *Confibina*, W.N.W. 12
From *Confibina*, to Port *Martin Lopes*, or, *el Condadillo*, W.N.W. 8
From Port *Martin Lopes*, to the River of St. *Michael*, West, 18
From the River of St. *Michael*, to that of *Guibaltique*, West, 3
From the River of *Guibaltique*, to that of *Linpa*, West, 3
From the River of *Linpa*, to *Sacatecolula*, West, 4
From *Sacatecolula*, to the burning Mountain *San Salvador*, West, 10
From *San Salvador*, to *Bernal*, West, 4
From *Bernal*, to the Hill of *Italias*, N.W. 3
From the Hill of *Italias*, to the Port *Catulta*, or, of *Sonsonate*, N.W. 3
From *Sonsonate*, to the River of *Moticalco*, N.W. 7
From the River of *Moticalco*, to the Bar of *Estapa*, West, 19
From *Estapa*, to *Sapotitlan* burning Mountain, N.W. 12
From *Sapotitlan*, to *Milpas*, two burning Mountains, N.W. 12
From *Milpas*, to *Soconusco* burning Mountain, N.W. 12
From *Soconusco*, to la *Encomienda*, or, the Cross, West, 7
From la *Encomienda*, to *Bernal*, N.W. 5
From *Bernal*, to Port *Mosquitos*, N.W. 9
From Port *Mosquitos*, to *Teguantepeque*, N.W. 8

From *Teguantepeque*, to *la Ventofa*, W. N. W. 4
From *Teguantepeque*, to *las Salinas*, or the Salt-Pits, *W. N. W.* 6
From *Salinas*, to the Head-land of *Macatan*, N. W. 2
From *Macatan*, to the Head-land of *Banba*, W. N. W. 4
From the Head-land of *Banba*, to the Ifland of *Itata*, *Weft*, 3
From the Ifland of *Itata*, to Point *Artea*. W. N. W. 7
From Point *Artea*, to the River of *Samitlan*, N. W. 4
From *Samitlan*, to the River of *Copalita*, *Weft*, 2
From the River of *Copalita*, to *Guatulco*, *Weft*, 1
From *Guatulco*, to the Ifland of Sacrifices, *Weft*, 2
From the Ifland of Sacrifices, to the River *Coiula*, N. W. 3
From the River *Coiula*, to that of *Julian Carrafco*, *Weft*, 4
From the River of *Julian Carrafco*, to Port Angels, *Weft*, 5
From Port Angels, to the River *de la Galera*, W. N. W. 3
From the River *de la Galera*, to the River *Mafia*, W. N. W. 18
From the River *Mafia*, to *Puerto Efcondido*, or, hidden Harbour, *Weft*, 10
From *Puerto Efcondido*, to *Pefqueria de Don Garcia*, W. N. W. 30
From *Don Garcia's* Fifhery, to Port *Marques*, N. W. by W. 5
From Port *Marques*, to *Acapulco*, 1
From the Port of *Acapulco*, to *Tequepa*, N. W. 18
From Point *Tequepa*, to the Head-land of *Petaplan*, N. W. 10
From *Petaplan*, to Port *Siguatanejo*, N. 4
From Port *Siguatanejo*, to *Tacomatlan*, or, *Iftapa*, N. W. 2
From *Iftapa*, to *Sacatula*, N. W. 12
From *Sacatula*, to *Tutapan*, N. W. ½

From

From *Tutapan*, to *Muibuta*, or *Pomaro*, N.W.	4
From *Pomaro*, to *Colima*, or *Santiago*, N.W.	8
From *Santiago*, to the Port of the *Nativity*, and that of *Melaque*, N.W.	6
From Port *Melaque*, to the Rocks of *Aquiapasulco*, N.W.	7
From the Rocks of *Aquiapasulco*, to the Islands of *Chametlan*, N.W.	3
From the Islands of *Chametla*, to Cape *Corrientes*, or Currents, N.W.	10
From Cape *Corrientes*, to Point *Ponteque*, N. by E.	10
From Point *Ponteque*, to the Islands *Tres Marias*, N.W. by W.	17
From *Ponteque*, to Port *Matanchel*, or *Xalisco*, N.E.	20

CHAP. IV.

Of the Winds and Currents in the South Sea; as also a large Table of the Latitudes and Longitudes of all remarkable Places along that Coast.

Suppose an imaginary Line from Port St. *Mark* at *Arica*, to Point *Aguja*, or, Needle-Point, which is near the Port of *Paita*, drawn 30 Leagues at Sea from each of those Ports, from that Line to the Coast, the S.E. and S.S.E. Winds reign all the Year. In Winter they are very boisterous, and keep more to the S.E. But it is to be observ'd, that within a League or two of the Coast, there are sometimes *North* and *N.E.* Winds, which are not very lasting, and blow weekly, and are most frequent in the large open Bays along the Coast.

Suppose another imaginary Line from the said Point *Aguja*, or, Neddle-Point, to Point *Santa Elena*, 20
Leagues

Leagues out at Sea from each of them, and from that to the Continent, the *South* Wind reigns all the Year; but five or six Leagues from the Shore there are sometimes *S. W.* Winds, more especially in the Angles the said Coast makes, and these Winds are generally moderate, but not lasting.

Imagine another Line ten Leagues out at Sea from the said Point *Santa Elena*, to Cape *Pasado*, and between that and the Shore the Wind is *S. W.* all the Year.

From Cape *Pasado*, to Cape St. *Francis*, draw also a Line five Leagues out at Sea, and between it and the Land the *S. W.* Wind prevails. Not to limit the several Winds which happen to blow without these Lines, according to the Times of the Day or Night, sometimes coming off the Shore, and then again from the Sea, and being more or less boisterous, according to the Season of the Year.

In like Manner, draw a Line from Cape St. *Francis*, to *Morro de Puercos*, or, the Head-land of Swine, and all to the *Eastward* of that is call'd *La Travesia*, that is, the crossing to *Panama*; and here is a Winter and Summer Season, tho' after an odd Manner, that is, without Regard to the Nearness of the Sun; for, according to the Course of Nature, the Summer ought to begin there on the 25th of *March*, when the Sun passes the Equinoctial, to the *Northward*, on which Side that Coast and Sea lie, where he should produce the usual Effects 'till the 25th of *September*, when that noble Planet crosses the Equinoctial again to the *Southward.* Yet is this known to be otherwise; for the Summer along this cross Sea and Coast of *Panama*, begins in *January*, when the Sun is farthest to the Southward of the Equinoctial, so that there the Season is contrary to the Course of Nature, and the Winter begins in *June*, when the Sun is on the *North* Side, which is directly opposite to the Effects of the Sun.

Along

Along this Coast of *Panama*, and the Sea before it, there are six Summer and six Winter Months. The Summer begins in *January*, and ends in *June*, and during this Season the *North*, *N. E.* and *N. W.* Winds reign. In *January*, *February*, and *March*, they are very boisterous, and there falls no Rain along that Coast of *Panama*, Port *Pinas*, *Malpelo*, *Puerto Quemado*, or, Burnt Haven, and all the rest on, as far as Cape St. *Francis*. At the same Time it rains much on the Coast of *Manta* and *Guayaquil*; and the Reason is, because those reigning Winds have drove the Clouds upon that Coast, and the said Winds stopping there, the Clouds can pass no farther, but are dissolv'd by the Sun, and fall down in heavy Showers. These reigning Winds, during the three first Summer Months, sometimes reach as far as *Manta*, Point *Santa Elena*, and Cape *Blanco*; and sometimes they do not reach to Cape St. *Francis*, which happens according as they are stronger or weaker on the Coast of *Panama*.

During these same three Months, there is generally an *E. N. E.* Wind reigning about *Malpelo*, being a settled Breeze and fair; and between *Malpelo* and the Land of *Buenaventura*, this Wind becomes *North*; and from within Sight of the Island *Gorgona*, to *Puerto Quemado*, or, Burnt Haven, it is generally *N. W. W. N. W.* and *West*, with heavy Showers of Rain.

Such is the Weather during these three Summer Months, and such is the Variety of it, according to those several Places. About the first Days of *April*, the Rains begin to fall all along that Gulph and Coast of *Panama*, and the peaceable Winds prevail, with Calms, for the most Part, or those they call *Virazones*, which are *South*, *S. W.* and *S. S. W.* Winds, and sometimes they fly over to *N. W.* generally with most violent heavy Rains; and thus the Winds flutter between strong Gusts, gentle Gales, and Calms, 'till the End of *June*, when the Summer ends.

In *July* begin those they call *Vendavales*, and last 'till

'till the End of *December*, and thefe *Vendavales* are *S.* and *S.W.* Winds, with mighty Rains, Thunder, and Lightning, and the Fury of thefe *Vendavales* is in *September*, *October*, and *November*; and then at Times about *Panama*, the *S.W.* Winds blow up fair Weather for ten or twelve Days together, and they are not fo fierce as to obftruct Navigation. They blow generally moft fierce during the aforefaid fix Winter Months, and fometimes the Wind will flip over to the *N.* and *N. E.* with heavy Rains; but that lafts not long, and it never reaches above 20 Leagues out at Sea.

During the fame Seafon, there are fometimes *Weft* and *W. S. W.* Winds, which carry over the Ships to the Coaft of *Peru*, and at Night the Wind ufes to come about *N. W.* with heavy Showers; but that is not lafting, which may be met with half Way over the Gulph, to *Panama*. When the *Northerly* fettled Winds reign at *Panama*, there are ufually Calms and good ftill Weather from Cape St. *Francis*, to Cape *Blanco*; and when the Summer begins at *Panama*, then the Winter commences at *Guayaquil*, and it rains five Months in the Year, that is, from the Beginning of *January* 'till the End of *May*; and all the Winds blow from the Ifland *Santa Clara* towards the River, and it thunders and lightens very much, more efpecially on the Mountains of *Cuenca*, which are thofe that appear on the right Hand going up the River; and yet at the fame Time, for the moft Part within the River, the Weather is fair and calm. Here the Summer commences in *June*, when it does not rain; but the *Weft* Wind blows very hard, which the Natives call *Chanduy*.

Cape *Blanco* is very pleafant and calm for four Months in the Year, which are *January*, *February*, *March*, and *April*; all the reft of the Year is very ftormy, and the Current fets up from the faid Cape, that is, to the *Southward*.

It is abfolutely neceffary in Navigation, to be acquainted with the fetting of the Currents, which frequently

quently put Ships from their intended Courſe, and run ſo unperceivably, that when a Pilot expects to make one Land, he finds himſelf upon another he never thought of, occaſion'd by the Force and Rapidity of the Waters ſetting along the Coaſts and ſpacious Bays ; nor is he able to perceive which Way they drive, 'till he makes the Land, and by it diſcovers whether the Ship has made a long or a ſhort Run. This he may alſo diſcover by the Latitude, and his Approach to, or Diſtance from the Equinoctial, and ſo find out, whether he has advanc'd much or little, in Proportion to the Wind as it was for or againſt him, and by that the Courſe of the Water will appear.

This ſetting of the Currents, has occaſion'd the Loſs of many Ships ; for ſometimes they run towards the *Eaſt*, or the *Weſt*, or *South Weſt*, in ſuch Places as is unknown to Sailors ; and thus they are often a-ground on ſome Bank or Shoal, without any Fault on their Side, as being altogether unacquainted with that Motion of the Water. In other Places the Water has no Motion at all, and it is ſafer in Sailing, to ſuſpect the Motion of the Water, than to rely too much upon it, and particularly when the Sun is not to be ſeen, and conſequently the Latitude cannot be taken ; for it may happen, that when the Ship ſtands her due Courſe, the Pilot may fancy the Water ſets ſome other Way, and conſequently make an Allowance for it, by which Means he will in the End find himſelf out in his Reckoning; for the Water is ſo uncertain, that there often appear on it Tokens of ſtrong Currents, Streams, or other Motions, and at the ſame Time it is quite ſtill, which is made out by the Land. However, great Regard muſt be had to obſerve Currents, eſpecially in Sailing, where any Effects of them appear, always providing with Caution, leſt they force the Ship aſhore, or upon ſome Bank of Sand, or imbay it ; to prevent which, there is to be timely Allowance, that the Danger may be prevented.

As

As soon as the Sun is gone over to the *South* of the Equinoctial, which commences the Summer in the *Southern* Parts, the Waters begin their Motion, setting *South* and *S.W.* and this they do from Cape St. *Francis* along the Coast, and thirty or forty Leagues out to Sea; and in the same Manner, when the Sun crosses the Equinoctial, to the *Northward*, they move back, and begin to run to the *N.* and *N. W.* from the Port of St. *Mark* at *Arica*, along all the said Coasts, and for the same Breadth of thirty or forty Leagues out at Sea. And it is to be obserʋ'd, that in all these Motions, either in Summer or Winter, they always bend off from the Coast; and this is the most general Rule in those Parts, tho' there are some Exceptions, and the contrary happens in some Places; but that does not hold for any long Space, and you soon come again into the usual Course.

From the aforesaid Cape St. *Francis*, as far as *Malpelo*, it is most certain the Current sets *E.* and *E. S. E.* towards the Island *Gorgona*, and the Bay of *Buenaventura*; and this is most frequent in Winter, tho' sometimes the Water is quite still.

From *Malpelo*, as far as the Head-land *Morro de Puercos*, the Water has generally no Run at all.

From the Island *Gorgona*, to Cape St. *Francis*, the Current seldom sets to the *S. W.* but its usual Motion is to the *N. W.* and at other Times it stands still.

From the Island *Gorgona*, to the Head-land *Morro de Puercos*, the Water sets along the Coast Winter and Summer towards the *N. W.*

When the Trade-Winds prevail, the Sea between *Morro de Puercos* and *Malpelo* sets towards the *S. W.*

The setting of the Sea and Currents is so various, that no human Understanding is able to comprehend it, not only in these Seas, but in many other Parts of the World. It is a Secret Providence has conceal'd from us. Experience informs us of its Being; but the Reasons of it are above our Reach.

A

The South Sea *Coasts.*

A TABLE

OF THE

Latitudes and Longitudes

OF

All the noted Ports, Islands, Rivers, Bays, Capes, and other Places worth observing along the *Western* Coast of *America*, that is, in the *South Sea*, from *California* in the *North*, to the Streights of *Magellan* in the *South*, from the same Manuscript *Spanish* Coasting-Pilots, as above.

Note, *That the Longitudes are taken from the Westermost Point of the Island* Gran Canaria, *the largest of the Canaries, where the Spaniards generally place their first Meridian.*

	Latit. D. M.	Longit. D. M.
THE Island of *California*,	24 40	255 15
The *Eastermost* Head of it,	24 40	258 15
The Point,	25 30	259 50
Cape St. *Lucas*,	25 42	259 07
The *Southermost* Part,	24 40	285 00
Last Point of the Continent,	24 40	260 55
The River *de la Sal*, or, of Salt,	23 30	262 10
Las Chamitas,	22 55	262 48
The River of St. *Andrew*,	22 30	264 08

The

A Description of

	Latit.		Longit.	
	D.	M.	D.	M.
The Islands *Tres Marias*,	22	07	264	14
The River of *San Milpa*,	22	05	264	23
Boca de las Higueras,	21	32	264	38
Punta de la Cruz, or, Point of the Cross,	21	26	264	16
The Island of *Calisto*,	20	10	264	22
Cape *Corrientes*, or, Currents,	20	20	265	20
Juan Ballegas,	20	28	265	50
Cabo de los Angeles, or, Cape Angels,	20	20	266	00
New Galicia,	20	25	266	26
Puerto de la Navidad, or, Port Nativity,	20	10	266	40
Bay of *Santiago*,	20	04	266	08
River of St. *Peter*,	19	52	267	30
River of *Aculima*,	19	30	267	50
River of *Sacatula*,	18	40	269	16
Island *de Ladrillos*, or, of Bricks,	17	52	270	05
River *de Gaviotas*,	17	40	270	28
Farallon, or, the Rock,	17	35	270	16
Point *Siguantanejo*,	17	20	270	04
River *Piticalla*,	17	15	270	55
River of *Mitala*,	17	08	271	28
River of *Sitala*,	17	40	272	04
Port of *Acapulco*,	17	00	272	04
Rio de Pescadores, or, Fisher-mens River,	17	00	272	45
Rio de Don Garcia, or, Don Garcia's Riv.	16	45	273	00
Punta de la Galera, or, the Galley Point,	16	08	273	42
Rio Verde, or, green River,	16	08	273	45
The Hill of *Talcamanca*,	16	00	273	55
Puerto Escondido, or, hidden Port,	15	50	274	32
The Island *de la Brea*, or, of Tar,	15	40	274	45
River *Milcas*,	15	38	275	00
River *de la Galera*, or, of the Galley,	15	36	276	06
Port *Angeles*, or, Angels,	15	26	276	06
River *Carrasco*,	15	18	276	18
River *Dicilo*,	15	20	276	40
Port *Aguatulco*,	15	36	276	25
The Head-land of *Masatatlan*,	15	30	277	46

The

The South Sea Coasts.

	Latit.		Longit.	
	D.	M.	D.	M.
The Island *Itata*,	15	30	277	26
Las Salinas, or, the Salt-pits,	15	42	278	20
The Bay of *Teguantepeque*,	15	50	278	46
Barra de Macias,	15	20	278	46
Morro, or, the Head-land *Bernal*,	14	56	279	47
Cerro de la Encomienda, or, Cross Hill,	14	58	280	00
The burning Mountain of *Soconusco*,	14	51	280	36
Bay of *Milpas*,	14	51	281	07
River of *Anabasos*,	14	29	282	20
River of *Sapotitlan*,	14	40	281	49
The Bar of *Istapa*,	14	24	282	50
Rio Grande, or, the great River,	14	20	283	40
River *Motualpe*,	14	07	284	00
Port of *Sonsonate*,	14	00	284	35
Point *Remedios*,	13	48	284	38
The Head-land of *Icacos*,	13	55	285	00
The Coast of *Tonela*,	13	50	285	22
Bar of *San Salvador*, or St. *Savior*,	13	40	285	55
The River of *Lampa*,	13	10	286	30
The River of St. *Michael*,	12	45	287	05
The Bay of *Condadillo*, or, the lit. County	12	38	287	46
Punta Gorda,	12	30	287	45
The Gulph of *Amapala*,	12	20	288	08
Point *Arenas*,	12	10	288	10
Port *Realejo*,	12	30	288	48
Punta del Leste, or, *East* Point,	11	40	289	00
The Bay of *Tosta*,	11	30	290	10
Gulph *del Papapayo*, or, of the Parrot,	11	10	290	37
Point St. *Catherine*,	10	34	288	48
Port *Delas*,	10	30	289	00
Morro Hermoso, or, Beautiful Cape,	09	17	290	10
Point *Guiones*,	09	47	290	17
Cape *Blanco*, or, White Cape of *Nicoya*,	09	20	290	50
Morro de la Ensenada, or, H. la. of the Ba.	10	10	291	20
The Bay of *Nicoya*,	09	18	291	49
The *East* Point of *Caldera*, or, the Kettle,	09	50	291	56

Y   Port

A Description of

	Latit.		Longit.	
	D.	M.	D.	M.
Port *Caldera*,	09	43	292	27
La *Herradura*, or, the Horse-shoe,	09	20	292	40
Rio de la *Estrella*, or, River of the Star,	09	08	292	47
Puerto del *Ingles*, the *English*-man's Port,	09	00	293	00
Punta *Mala*, or, bad Point,	08	55	293	17
The Island *del Cano*,	08	45	293	30
Golfo *Dulce*, or, fresh Water Gulf,	08	47	293	05
Point *Burica*,	08	30	294	21
Port *Limones*, or Lemons,	08	38	294	16
Island de *Limones*, or, of Lemons,	08	17	294	10
River of *Chiriqui*,	08	37	295	00
La *Montuosa*,	08	53	295	36
Pueblo *Nuevo*, or, new Town,	07	22	295	40
The Island of *Quicara*,	07	41	295	00
Point *Philipinas*,	07	40	296	10
Point *Mariato*,	08	26	296	52
Bay of *Philipinas*,	07	12	296	40
The Head-land of *Riercos*,	07	15	296	50
Point *Higuera*, or, of the Fig-Tree,	07	21	297	44
Punta *Mala*, or, bad Point,	07	35	297	07
River *Mensave*,	08	47	297	40
River *Covita*,	08	01	298	35
River of *Parita*,	08	11	298	36
River of *Nata*,	08	26	298	37
Port of *Villa*,	08	28	299	58
Portete,	08	56	299	20
River *Caymito*,	09	09	299	30
Island *Otoque*,	08	30	299	37
Island *Taboga*,	08	40	299	40
Ancon,	08	55	299	50
Panama City,	09	00	300	56
Chepillo,	09	00	301	01
Point *Manglares*, or Mangroves,	08	53	300	23
Island *Contadora*,	08	46	300	32
Isla del Rey, or, King's Island,	08	10	300	05
Cape St. *Laurence*,	08	10	300	58

River

The South Sea Coasts.

	Latit.		Longit.	
	D.	M.	D.	M.
River *Congo*,	07	53	301	43
St. *Michael's* Bay,	08	18	301	20
Point *Garachine*,	08	10	301	30
Point *Caracoles*,	07	52	301	00
Point *Pinas*,	07	24	301	08
Morro Quemado, or, burnt Head-land,	06	45	301	19
Puerto Claro, Port clear,	06	46	301	37
Bay of St. *Francis*,	05	50	301	50
Bay of St. *Anthony*,	06	20	302	00
Port of *Indians*,	06	14	302	02
Watering Places and Coasts of *Anegadas*,	06	55	302	03
River *Sandi*,	05	35	302	05
Island of *Coco's*,	05	09	299	08
Las Salinas, or, the Salt-pits,	05	15	302	05
Cape *Corrientes*, or Currents,	05	00	301	57
River of the *Noamas*,	04	38	302	23
Bay of *Buena Ventura*, or good Fortune,	04	08	302	41
River of *Buena Ventura*,	04	00	302	50
Island of *Malpelo*,	04	00	299	46
River *Pisco*,	03	45	302	39
River of the *Magdalen*,	03	35	302	20
River of Cedars,	03	25	302	07
River *Gorgon*,	03	24	301	54
Island *Gorgona*,	03	15	301	36
Head-land of *Barbacoas*,	02	20	300	45
Island *del Gallo*,	02	17	300	40
Bay of *Mira*,	02	05	300	50
River of *Mira*,	01	57	300	26
Island *Gorgonilla*,	01	58	300	25
Point *Manglares*, or Mangroves,	01	40	300	10
Ancon de Sardinas, or little Pilchard Bay,	01	15	300	02
River of *Santiago*,	01	14	299	30
Cape St. *Francis*,	00	50	299	57
Portete,	00	46	299	05
Cojimies,	00	25	299	00
River *Juma*,	00	05	298	44
Cape *Pasado* to the *Southward*,	00	08	298	32

A Description of

	Latit.		Longit.	
	D.	M.	D.	M.
Bay of *Caracas*,	00	28	298	43
Bay of *Manta*,	00	50	298	31
Cape St. *Laurence*,	01	09	298	10
Island *Plata*, or, of Plate,	01	15	298	15
Bay of *Salango*,	01	20	298	56
Los *Ahorcados*,	01	25	298	15
Island of *Salango*,	01	40	298	25
River *Colonche*,	02	00	298	18
Point St. *Elena*,	02	20	298	04
Point *Carnero*,	02	26	298	10
River of *Chanduy*,	02	27	298	38
Point *Chanduy*,	02	56	298	43
Boca *Chica*,	02	40	299	00
Bay of *Chanduy*,	02	26	299	14
The Point of the Bay,	02	43	299	24
Island *Puna*,	02	54	299	10
Island *Santa Clara*,	03	23	298	50
Isla Verde, or, Green Island,	02	26	299	48
River of *Bolao*,	03	36	298	38
River *del Buey*, or, Ox River,	03	40	299	20
River of *Payana*,	03	50	298	50
River of *Tumbes*,	03	52	298	40
Point *Mero*,	03	54	298	25
Mancora,	04	10	298	17
Cabo Blanco, or, White Cape,	04	22	298	06
Point *Farina*,	04	50	298	06
Bay of *Colan*,	04	55	298	26
Point *Paita*,	05	17	298	17
Little Island of *Lobos de Paita*,	05	23	298	40
Bay of *Sechura*,	05	23	299	00
Point *Nonura*,	05	45	298	40
Point *Aguja*, or, Needle Point,	06	00	298	06
Island *Lobos*, next the Main,	06	25	298	24
Head-land of *Eten*,	06	08	298	45
Island *Lobos* to Seaward,	06	33	298	10
River of *Sana*,	06	40	299	37

Head-land

The South Sea Coasts.

	Latit.		Longit.	
	D.	M.	D.	M.
Head-land of *Requen*,	06	28	299	17
Port *Cheripe*,	07	00	299	50
River *Pacasmayo*,	07	18	300	07
Port *Malabrigo*,	07	30	300	18
Head-land of *Malabrigo*,	07	47	300	10
Bay of *Cao*,	07	44	300	37
Port *Guanchaco*,	08	00	300	50
Head-land of *Carretas*,	08	15	300	50
Head-land of *Guanape*,	08	34	301	02
Rocks of *Guanape*,	08	36	300	50
Port *Santa*,	09	00	301	15
Island *Santa*,	09	10	301	10
Port *Guambacho*,	09	20	301	20
Port *Casma*,	09	28	301	30
Port *Bermejo*, or, Red Port	09	40	301	38
Guarmey,	10	00	301	45
Port and River *de la Barranca*,	10	30	301	46
Port *Guaura*,	10	52	301	43
Head-land of *Carquin*,	11	00	301	37
Marsoque,	11	05	301	30
Malgesi,	11	08	301	28
Las Perdices, or, the *Partridges*,	11	16	301	52
Port *Chancay*,	11	32	301	51
Los Pescadores, or, the Fisher-men,	11	46	301	56
Las Ormigas, or, the Pismires,	11	55	301	35
Port *Callao*,	12	06	302	15
Port *Chilca*,	12	20	302	25
Island of *Asia*,	12	35	302	14
Port *Canete*,	13	02	302	37
Port *Chincha*,	13	34	302	55
Port *Pisco*,	13	50	303	00
Island *Sangalla*,	14	05	302	35
Morro de Viejas, or, old Wom. Head-land,	14	22	303	15
Island *Lobos*,	14	28	303	20
Morro Quemado, or, Burnt Head-land,	14	30	303	34
River *Tea*,	14	44	303	33

Y 3 Head-land

A Description of

	Latit. D. M.	Longit. D. M.
Head-land of *Canas*, or, of Reeds,	15 00	303 50
Port St. *Nicholas*,	15 06	304 40
Port St. *John*,	15 15	304 15
Bay of *Arequipa*,	15 25	304 40
Point *Arequipa*,	15 37	304 36
Port *Chala*,	15 40	305 00
Head-land of *Atico*,	15 48	305 46
River of *Camana*,	15 46	306 19
Port of *Ocana*,	15 57	306 32
Creek of *Camana*,	16 20	307 35
Creek of *Chilca*,	16 30	307 50
Island *Guano*,	16 40	308 09
The *Guaca* and Point of *Cornejo*,	16 48	308 15
Point *Ilay*,	16 48	308 45
Head-land of *Ilay*,	17 07	308 40
Creek of *Chala*,	17 00	309 03
River *Tanbo*,	17 02	309 15
Point *Yerba Buena*,	17 40	309 35
Port *Ilo*,	17 42	309 40
Point *Ilo*, and its Rocks,	17 46	309 35
Head-land of *Sama*,	17 55	310 19
River of *Quiaca*,	17 45	310 35
Port of *Arica*,	18 00	311 08
Head-land of *Arica*,	18 05	310 55
The Break of *Vitor*,	18 30	311 08
Break of *Camarones*,	19 17	311 14
Break of *Pisagua*,	19 26	311 15
Head-land of *Tarapaca*,	20 00	311 05
Island *Iquique*,	20 00	311 52
River *Loa*,	21 06	311 17
Algodonales,	21 30	311 15
Port *Cobija*,	21 40	311 14
Point *Angama*,	21 54	311 05
The Bay to it,	22 06	311 09
Bay of *Mijillones*, or, Muscles,	22 00	311 15
Morro Moreno, or, Brown Head-land,	23 00	311 08

The

The South Sea Coasts. 327

	Latit.		Longit.	
	D.	M.	D.	M.
The Bay to it,	23	12	311	12
Morro de Jorge, or *George's* Head-land,	23	30	311	15
Bay of St. *Nicholas*,	24	20	311	25
Port *Betas*,	24	45	311	42
Point *Betas*,	24	54	311	30
El Juncal, or, the rushy Ground,	25	55	311	28
Chanalar, and its Islands,	25	14	311	26
Head-land of *Copiapo*,	26	58	311	14
Bahia Salada, or, Salt Bay,	27	05	311	10
Totoral,	27	50	311	19
Head-land del *Juncal*, or, of the rushy Gr.	28	36	311	03
Port *Guasco*,	28	30	311	03
Island *Paxaros*, or, of Birds,	29	46	310	10
Port *Coquimbo*,	30	00	310	40
Bay of *Longoi*,	30	24	310	46
Guanaquero,	30	16	310	40
Break of *Limari*,	31	00	310	35
Island of St. *Felix*, next the Continent,	26	15	303	15
The other to Seaward,	26	17	303	05
High-lands of *Chupa*,	31	14	310	37
River of *Conchali*,	31	26	310	50
Point *Ballena*, or, of the Whale,	31	32	310	50
Port *Guillermo*, or, *William*,	31	41	311	00
La Silla del Governador, or, the Gov. Saddle	32	00	310	00
Papudo,	32	25	311	29
River *Ligua*,	32	20	311	20
River *Quintero*, and the three Shoals,	32	28	311	26
River *Concon*,	32	45	311	23
Port *Valparaiso*,	33	00	311	30
Point *Curoama*,	33	10	311	10
Port St. *Antony*,	33	29	311	08
River *Maipo*,	33	43	311	03
The Shoals of *Rapel*,	33	50	310	57
Topocalma,	34	00	310	53
River *Maule*,	35	00	311	30
Point *Humos*, and its Shoals,	35	40	311	18

River

A Description of

	Latit.		Longit.	
	D.	M.	D.	M.
River of *Itata*,	36	05	311	27
Port of the *Conception*,	36	30	311	20
Island *Quiriquina*,	36	42	311	10
Point *Talcaguano*,	36	45	311	10
River *Biobio*,	36	54	311	00
Bay of *Aranco*,	37	12	311	12
Island of *John Fernandes*,	33	50	305	17
Island of *John Fernandes* to Seaward,	34	00	303	50
Island of *St. Mary*,	37	14	311	00
River of *Tucapel*,	37	48	311	03
River of *Imperial*,	38	27	311	06
Island *Mocha*,	38	28	310	46
River of *Tolten*,	39	12	311	21
Head-land of *Bonifacio*,	39	40	311	07
The Leeward Land of *Niebla*,	39	55	311	25
Valdivia,	40	00	311	10
Point *Gallera*, or, Gally Point,	40	00	310	43
Rio Bueno, or, Good River,	40	20	311	17
High-lands of St. *Peter*,	40	35	311	18
St. *Peter*'s Rock,	41	00	311	08
Point *Queda*,	41	30	311	08
Point *Godoy*,	41	45	311	23
Point *Caralmapo*,	41	52	311	30
Rocks of *Caralmapo*,	41	52	311	26
Point Whirlwinds at *Chacao*,	42	00	311	18
Point of the Cross to Leew. of the Island,	42	12	000	00
Tetas, or Dugs of *Cucao*,	43	00	311	02
Point *Cilan*, to Windw. of the Isl. *Olygue*,	42	00	311	00
Island *Guafo*,	44	20	310	46
Corcobado on the Continent,	43	30	313	28
Point of the Bay,	44	05	313	00
Windward Point of the Bay,	44	05	313	00
Cape *Corzo*,	46	35	312	22

The *Spanish* Manuscript goes no farther *South*, this being the utmost Extent of their Trade.

FINIS.

The INDEX.

A

A Caguna Hill, page 222
Acapulco Prize, vid. Batchelor Frigat.
Acapulco Port, 300, 301, 302, 303, 304, 307
Acari, 214, 215
Aguja Point, 152, 153
Ahorcados Rocks, 141
Alcatrezes Island, 300
Algodonales, 232
A cwance at Sea short, 3
Alquivire, 244, 245
Amargos Port, 248
Ancon de Rodas, 191
Ancon de Sardinas, 129
Ancon fin falida, 251
Anco Point, 250
Andalin River, 240
Andalita River, 242
Anegadas Islands, 117, 118, 119
Angels Port, 296
Apufiguale Strand, 304
Aquapafulco, 308
Areca Tree, 21
Arena Point, 144
Arequipa, 215, 217, 219
Arica, 222, 225, 226
Arrival at *Guam*, or the *Ladrones*, 5
  At *Buton* Island, 42
  At *Batavia*, 52
  At the Cape of *Good Hope*, 66
  In the *Texel*, 99
  In the *Thames*, 108
Artea Point, 297, 298
Ascension Island, 94
Afferradores Creek, 280
Afia Island, 202
Atacama, 232
Atico Head-land, 216, 217

B

Baco Island, 263
Baia de Nuestra Senora, 234
Baia Honda, 264, 265
Baia Salada, 235
Ballena Point, 135
Ballefta Island, 204, 207
Balfam Coast, 287
Bimba Head-land, 297
Bantam Point, 63
Barbacoas River, 125, 126
Barca Port, 187, 211
Barranca River, 180, 183, 184
Barrancas Vermejas, 134, 197
Batavia Bay represented, 51
  Its Road, 53
  The City, 54
Bayan River, 110
Bearings of Islands in *India*, 35
  Of *Gilolo*, 38
  Of other Islands, 39
  Of *Bouro* Island, 40
  Of *Cambava* and *Wanfhut*, 42
  Of *Solayo* and *Celebes*, 46
  Of *Japara* and *Carimon Java*, 50
Bernal and Bernallillo, 287, 294
Betas Port, 234, 235
Biobio River and Dugs, 243
Bite of *Gilolo*, 34
Blanco, or, white Cape, 147, 148, 149, 262, 271, 272, 273

Boca-

The INDEX.

Bocatuerta River, 112
Bonbacho Port, 172, 173
Bonbacho burning Mountain, 274
Bonifacio Head-land, 245, 248
Boqueron, 199
Bora Island, 97
Bouro Island, 39
  Bearing and Description of it, 40
Bragueta, 294, 295
Bread Tree, or Rima, 21
Buenaventura Bay, 119, 120, 122
Buenaventura River, 121
Bufadero, 298
Buffin's Island, 63
Buracas Indians, 269
Burica Point, 262, 268
Buton Island describ'd, 42
  The Town of the same Name, 43
  Cut of the Island, and Ports about it, 45

C

Caballa Point, 210, 211, 212
Callacalla River, 248
Callao Island, 189, 191, 192, 193, 194, 194, 199
Callao Town, 195
Callo Port, 140
Camana, 217
Camarones Break, 226, 227
Camas Point, 136
Cambava Island, 41
  Bearing of it, 42
Canales Island, 264, 266
Canete, 202, 203
Cano Island, 262, 269, 271
Cape of Good Hope, 66
  Description of it, 67
Caracas Bay, 136
Carascoles Port, 114
Cavalnapo Fort, 250
Caravaillo River, 195
Carimon Java, how it shews, 50
Carnero Point, 142
Carnero Port, 243
Casma, 173, 174
Castro de Chiloe Town, 251

Catulta Strand, 290
Caucato, 204
Cechura Bay, 152
Cedros River, 125
Ceram Island, 39
Certificates of civil Entertainment between English and Spaniards, 11
  Of the Surgeons, for leaving a sick Spaniard at Guam, 12
Chame Head-land, 257
Chametla Islands, 308
Chancai, 190, 191
Chancaillo, ibid.
Chandui, 143
Chao Head-land, 166, 167
Chepillo Island, 110, 196
Chepo, 110
Cheripe, 155, 158, 159, 161, 163
Cheveral, 235
Chilca Point, 201, 202
Chile River, 238
Chiloe Island, 250
Chiman River and Island, 112, 196
Chinbote, 171
Chincha, 203, 207
Chinilla River, 111
Chira Island, 270, 271
Chiriqui Island and River, 266, 267, 268
Choapa, 237
Chola Port, 215
Choropoto Crags, 136
Chuche Island, 111
Chule Port, 218
Cipanso River, 270
Claro River, 248
Coiba Island, 263, 264
Cojimies Rivers, 134, 135
Coñula River, 298
Colan River, 150
Colanche River and Island, 141
Coles Indians, 269
Colima, 306, 307
Committee resolves to sail from Guam for India, 12
  Resolves for Talno, Ternate, or Mindanao, 33

Takes

The INDEX.

Takes Care of Goods and Provisions, 37
Orders Money to purchase Plunder, 61
Orders what Persons shall be in each Ship, ibid.
Resolves for the Cape of Good Hope, 65
Compostela City, 308
Conception Bay and Town, 241
Conception Port and Island, 240
Conchagua Island and River, 283, 284
Condadillo, 282
Consibina, 280, 281, 282
Constantina Island, 247, 248
Contreras Island, 266
Copalita River, 297
Copiapo Head-land, 234, 235
Coquinbo Port, 236
Corcobado Rock, 170
Coroama Point, 237, 239
Corral Port, 246, 248
Corrientes Cape, 117, 118, 119, 120, 308, 309
Corzo Cape, 251
Costa Rica, 269, 270
Courses and Distances from *Panama* to *Lima*, 196, and seq.
from *Panama* to *Acapulco*, 310
Craw-fish in the *South Sea*, 4
Cucao, 250
Curacancana, 212, 213, 214
Current setting *Northward*, 23
To the *Southward*, 25
Northward, 25
N. W. 37
Very strong, 38
North and South, 39
Westerly, 40
South, 47
See the Journal Tables in the *South Sea*, 316 & seq.

D

Day lost, or got, in sailing round the World, how, 52
Departure from *California*, 2
From *Guam*, 23
From *Buton*, 46

From *Batavia*, 62
From the Cape of *Good Hope*, 91
From the *Texel*, 108
Description of the Sea-coasts, Head-lands, Soundings, &c. along the *American* Coast, 109, & seq.
Distances of Places between *Panama* and *Lima*, 196
Dona Francisca Rock, 193
Don Martin Island, 185
Duarte Island, 63
Duke Ship, 2
Leaky, 72
Dutchess Ship, 2
Dutch Sailors run away, 102

E

Encomienda, 293, 294
Estapa Strand and River, 290, 291
Estrella River, 269, 270
Eten Headland, 154, 155, 158, 159
Extract of the Substance of the Journals, 103

F

Falara Port, 149
Favo Harbour, 271
Ferol Port, 172, 173
Fish-hook us'd by *Indians*, 22
Fonseca Bay, 282
Fraile Rock, 111
Frailes Islands, 258, 262

G

Galera Island and Point, 113, 132, 196, 248, 249
Galera River, 299
Gallo Island, 125, 128, 196
Garachine Point, 113, 114, 196
General's Island at *Batavia*, 51
Gilolo Island, 27
Bearings of the Point and Body of it, 38
Godoi Point, 250
Golfo Dulce Bay, 269
Gonzalo Head-land, 246, 248, 249
Goods sold for Provisions, 72
Gorgona Island, 122, 123, 124, 196
Gorgonilla Island, 129
Governadora Island, 262, 263
Governador

The INDEX.

Governador Port, 237
Gramadal Hill, 180, 183
Guacho, 187
Guadalupe, 161, 162
Guafo Island, 251
Guam Island, 5, 8
  Defcrib'd 13
  Cut of it, 16
Guanape. 164, 165, 166, 167, 169
Guanchaco, 164, 169
Guanico Mountains, 261, 262
Guano Island, 218
Guarco, 202
Guanmey Port, 177, 178, 179
Guafacalco River, 295
Guafco Port and Hill, 236
Guatimala, 290
Guatimala burning Mountains, 291
Guatulco, 296, 298
Guaura Port, 186
Guayaquil, 142, 146
Guayavas Island, 271
Guibaltique River, 285
Guiones Point, 273

H

Haguey, vid. Jaguei,
Herradura, 189, 240, 242, 269, 271
Hguera Point, 257, 258
Hormigas Island, 189, 191
Horn Island, near Batavia, 55
Hottentots, 69
Huevos Point, 240

I

Jaguei de la Cofta, 180, 181
Jaguei de Paquifa, 232
Jaguei de las Culebras, 176
Jaguei de la Zorra, 179
Japan, how it shews, 50
Java Island, 50, 51
Java Head, 64
Iguana, vid. Guam.
Iguanas Island, 257, 258
Ilai, 218
Ilo Port, 219, 220
Imperial River, 244, 245
India, the worft Country for Weather, 40
Ingles Port, 250

Journal Table from California, to the Island Guam, 3
  From Batavia, to the Cape of Good Hope, 66
  From the Cape of Good Hope, to the Northern Seas, 92
Isla Baxa, 309
Isla del Rey, 111, 112
Islands of India very numerous, 37
Islands of Solomon, 16
Islas de en medio, 271
Iftapa Town, 305
Italias Coast, 287, 288
Itata River, 240, 242
Itata Island, 297
Juan Diaz River, 110, 224
Julian Carrafco River, 298
Juncal, 235

K

King's Island, 111, 112, 248

L

Ladrones, or Marian Islands, 5, 14
  The Natives, 17
  Temperature, 18
  Port Umatta there, ibid.
Lagartos River, 111
Lago Bays, 250
Landecho Town, 270
Latitudes of Places in the South Sea, 319 and feq.
Lavapie, 243
Leon burning Mountain, 275
Leones Island, 262
Letter to the Governor of Guam, 6
  His Anfwer, 7
  Letter to the Owners, 73
  From the Owners, 100
Ligua Port, 237
Lima River and City, 195
Limavi, 237
Limones Point, 268
Limpa River, 286
Line of Battel, 88
Liquors taken in at Guam, 9
Lobos de Paita Island, 152, 153
Lobos Island, 186, 208, 209

Loma

The INDEX.

Loma Port,	214, 215	Mocha Island,	243, 244
Loma Quemada,	221, 222	Mongon,	174, 175
Longitudes of Places in the South Sea,	319 and seq.	Mongoncillo,	175, 176
		Monsons,	28
Longoi Bay,	237	Monte Christo Hill,	137, 138
Lora River,	232, 239	Montuosa,	262, 263
Lorinchincha,	204	Moratay Island,	26
		Morro de Jorge,	233

M

Macassar,	49	Morro del Carbon,	295, 296
Macatan Head-land,	296	Morro de Viejas,	207, 208, 209
Madalena de Cao River,	164	Morro de Puercos,	257, 258, 259, 260, 262
Madure,	49		
Maestra River,	111, 196	Morro Hermoso,	273
Magdalen Island,	251	Morro Moreno,	233
Magellan Streights,	252	Morro Quemado,	116, 117, 209
Mahe River,	112, 196	Morro Solar,	199, 200
Maire's Streights,	252	Mosquitos Port,	295
Malaca Bay,	149	Mota Island,	248
Malabrigo,	162, 163, 166, 167	Moticaleo River and burning Mountain,	290, 291
Mala Point,	202, 258		
Malpelo Island,	260	Motines,	305
Maltesi,	187, 188, 189	Mozupe Head-land,	157, 158
Mancora Hills,	147	Muibata,	306
Manglares, or Mangrove Point,	111, 129, 196		

N

Manila Ship, vid. Batchelor Frigat.		Nasca River,	212, 215
Manta Point,	136	Nata Head-land,	257
Marian Islands, vid. Ladrones.		Nativity Port,	304, 307
Mariato Point,	261, 262, 263	New Guinea,	34, 37
Mariquina River,	248	New Spain,	262
Marquis Ship,	2	Nicaragua,	262
Sold,	59	Niebla Point,	246, 248
Marquis Port,	300, 301	Nionimas River,	119
Marsoque,	187, 188	Noha Cape,	34
Martin Lopes Port,	282	How it shews,	35
Massa River,	299		

O

Matalotes Islands,	24	Ocana,	217
Maule River,	239, 240	Officer confin'd,	59
Melaque Port,	307	Officers supply'd,	59
Mero Point,	147	Allowances to them,	60
Mesa de Dona Mariana,	210, 212	Olleros, or, Potter's Point,	209, 210
Mesa de Roldan,	281	Onrest Island,	55
Miangola Island,	284	Ostiones River,	119, 120
Mijillones Bay,	232, 233	Otoqus,	111, 257
Milpas burning Mountains,	292, 293	Otoquillo,	257

P

Minas Hill,	121	Pacasmayo,	161, 162
Mindanao Island,	29, 30	Pachacama Rocks,	201
			Pacheca

The INDEX.

Pacheca Island, 111
Pajaros Islands, 235, 236
Paita, 150, 151, 152
Paitilla, 110
Palmas Island, 119, 121, 196
Panama, 110, 196, 257
Paneca burning Mountain, 288, 290
Papagayo Bay, 274
Papous Land, vid. New Guinea.
Papudo Port, 238
Paquifa Hill, 232
Paraca Port, 200, 205, 206, 207
Paramonga, or, Paramonguilla, 182
Paraos, a Sort of Boats, 19
Pariga River, 111
Parina Point, 149, 150
Passado Cape, 134, 135, 136
Passage between Solay and Celebes, 46
Paucora River, 110
Pena oradada, 52, 195, 199, 200
Pena oradada River, 112, 196
Penguin Island, 67
Pepper Bay, 65
Perico Port, 110, 196
Pero Lopez Creek, 121
Pescadores Rocks, 189, 191, 193, 217
Pesqueria de Don Garcia, 300
Petaplan Head-land, 304
Philipinas, 262, 263
Philippine Islands, 30
  Trade of them, 31
Pica Head-land, 231
Picoasa Mountains, 141
Piles Creek, 121
Pinas Port, 115, 196
Pine-Apple Tree, 22
Pisagua Break, 227, 228, 229
Pisco, 204, 205
Plata Island, 138, 139
Playa de las Perdices, 190
Plunder divided, 58
  Exchang'd, 60
Pomaro Town, 306
Ponteque Point, 309
Portete, 134, 196
Potocalma, 239

Prisoners set ashore at Guam, 13
Provisions taken in at Guam, 9
Pueblo Nuevo, 264, 265, 266
Puerto Bermejo, 176, 177
Puerto Chico, 300
Puerto Escondido, 299
Puerto Quemado, 117, 119, 196, 209
Puerto Seguro, 2
Puerto Viejo, 121
Pulababe Island, 63
Pulo Bouna, Pulo Shampo, and Pulo Cubina, 42
Puna Island, 142, 144, 145
Purification Town, 308

Q

Quedar Point, 249, 250
Quela Point, 250
Quenete, 202
Quiaca, 222, 223, 224
Quicara Island, 262, 263, 264
Quilca Creek, 217
Quintero Flats and Port, 238
Quiriquina Island and Port, 240, 241, 242

R

Realejo Port, &c. 275, 276, 277, 278, 280
Remedios Point, 287, 288, 290
Requen Head-land, 155, 156, 157
Returns made for Provisions at Guam, 10
Rima, the Bread-Tree, 21
Ring about the Sun 33
Rio Grande, 110
Rio Hondo, 112

S

Saavedra Islands, 24
Sacatecolula, 286, 287
Sacatula River, 305
Sacrificios Island, 298
Sailing Orders by the Dutch Admiral, 75
St. Catherine's Point, 273
St. Christopher's River, 248
St. Francis Cape, 132, 133, 134, 196
St. Helena Island, 92
St. John de Quaques, 134
St. John's Bay, 248
St.

The INDEX.

St. *John*'s Port, 207, 214, 274
St. *John*'s River, 121, 126
St. *Joseph*'s Shoal, 113
St. *Laurence*'s Cape, 112, 138, 139, 196
St. *Luke*'s Island, 271
St. *Mary*'s Island, 243, 252
St. *Matthew*'s Bay, 129, 130, 131, 196
St. *Michael*'s Bay, 113
St. *Michael*'s River, and burning Mountain, 284, 285, 287
St. *Nicholas*'s Port, 212, 213, 214
St. *Peter de Illoque*, 161
St. *Peter*'s Hills, 160
St. *Peter*'s Port, 249
St. *Vincent*'s Streights, 252
St. *Vincent*'s Port, 243
Salagua Port, 306, 307
Salango Island, 140
Salango Passage represented, 48
Salinas 187, 189, 296
Salinas del Piloto, 308
Salinas Point, 119
Sama Head-land, 221, 222, 223
Samitlan River, 295, 297
Sangallan, 204, 205, 206, 207
San Salvador burning Mountain, 286, 287
Santa Clara Island, 144, 147
Santa Elena Point, 138, 140, 141, 142
Santa Port, 170
Santiago Head-land, 306, 307
Santiago River, 129, 130, 191
Sapo Hill, 113, 115
Sapotitlan burning Mountain, 292
Sarate Island, 207
Sarpana Island, 15
Sebaco Island, 264
Secas, or dry Islands, 266
Sea-men supply'd with Money, 60
Sequetepeque, 292
Shetland, 97
Ships that came from the Cape of *Good Hope*, 76
Signals for keeping Company, 24
By Day, 76
By Night, 81

In a Fog, 85
For Ships to draw in a Line, 87
For chasing, 90
Siguatanejo, 304
Sitala Serand, ibid.
Soconusco burning Mountain, 293
Solayo Island, 46
Sonsonate, 288, 289, 290
Spaniards at *Guam* very obliging, 8
Spouts, 26
Supe, 184, 185

T

Taboga and *Taboguilla*, 111, 112
Tacames Bay, 131, 132, 196
Tacomatlan, 305
Tacoral burning Mountains, 227
Talao Island, 34
Talcaguano Port and Point, 240, 242, 243
Tanbillo, 119
Tanbo de las Perdices, 188, 190
Tanbo River, 219, 220
Tangolatango Island, 298
Tarapaca, 225, 229, 230
Teguantepeque, 276, 292, 295, 296
Telica burning Mountian, 276, 277
Tenlebi River, 128
Tequepa, 304
Ternate Island, 27
Tidore Island, ibid.
Tonela Coast, 286, 287
Tosta River, 275
Tosta Ridge, 276
Totoral, 235
Tres Marias Islands, 309
Tres Montes Cape, 251
Truxillo, 164, 168, 169
Tucapel, 244
Tuchisi Head-land, 306
Tuguman River, 110
Tumbes, 142, 147, 148
Tutapan Town, 305
Twelve Apostles Rocks, 252
Tygers, 64

V

Valdivia Town and Port, 248
Valparaiso Port, 237, 238
Variation

The INDEX.

Variation, none, 4
  Seven and nine Degrees, 5
  Five Degrees, 25
  See the Journal-Tables.
Vasa Borrachos, 134, 135, 197
Velas Port, 273
Ventosa, 295
Vieja Head-land, 200
Vitor Break, 226
Umatta Port, vid. Guam.
Volcan del Viejo, 227, 279, 282

W

Wanshut Island, 41
  Bearing of it, 42
Weather of *India* uncertain and unwholsome, 27, 29
Winds reigning in the *South Sea*, 313

Y

Yca, 207
Yellow-Tail, 22
Yerba Buena, 220

FINIS.

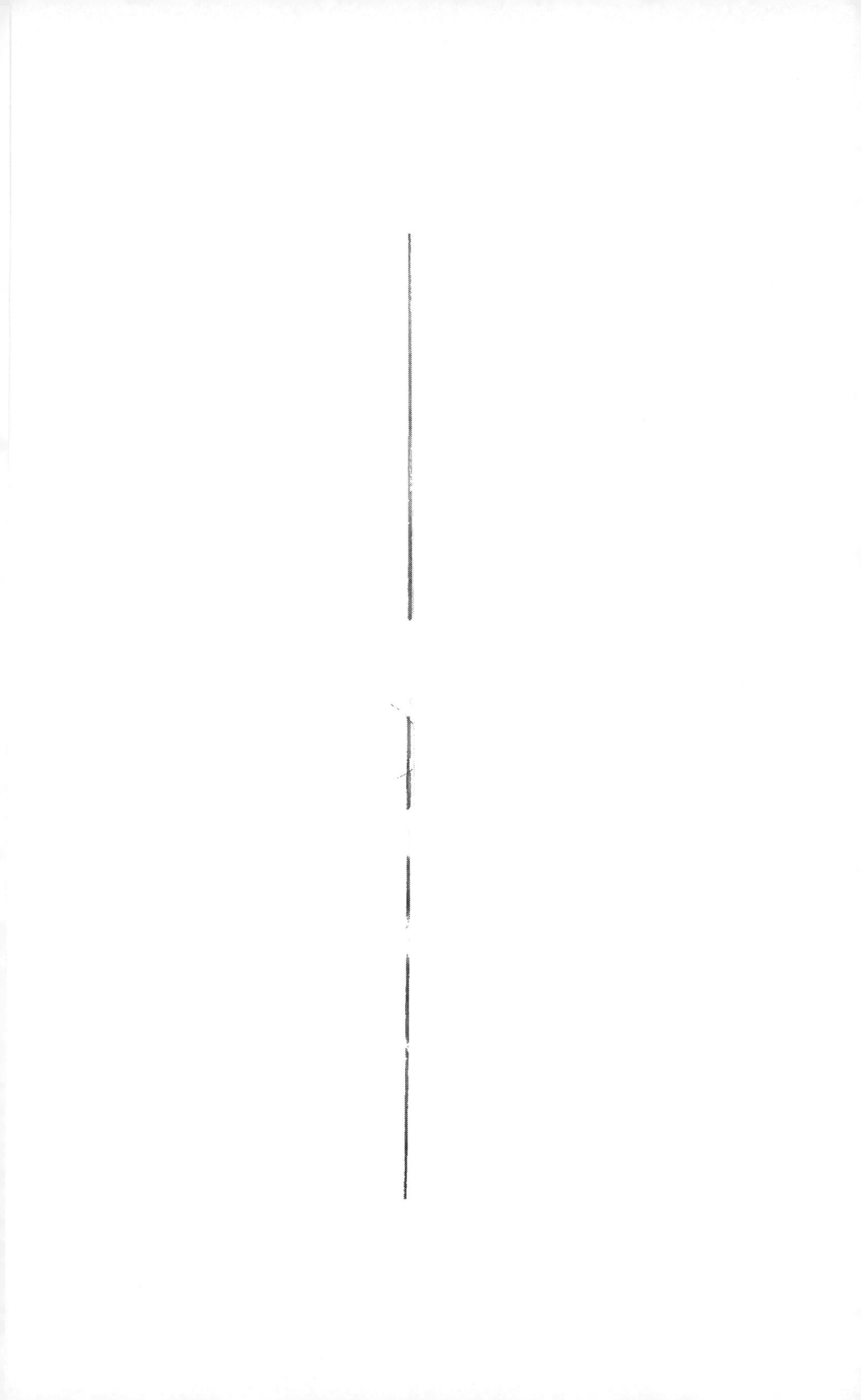

Lightning Source UK Ltd.
Milton Keynes UK
UKHW020640060223
416537UK00012B/2517